I0053137

One-Dimensional Transient Flow in Pipelines Modelling and Simulation

Authored by

Abdelaziz Ghodhbani
National Engineering School of sfax
University of Sfax
Tunisia, Africa

Ezzeddine Haj Taïeb
National Engineering School of sfax
University of Sfax
Tunisia, Africa

Mohsen Akrout
National Engineering School of sfax
University of Sfax
Tunisia, Africa

&

Sami Elaoud

National Engineering School of sfax
University of Sfax
Tunisia, Africa

One-Dimensional Transient Flow in Pipelines Modelling and Simulation

Authors: Abdelaziz Ghodhbani, Ezzeddine Haj Taïeb, Mohsen Akrout and Sami Elaoud

ISBN (Online): 978-981-5123-76-0

ISBN (Print): 978-981-5123-77-7

ISBN (Paperback): 978-981-5123-78-4

First published in 2023.

Bentham Science Publishers Pte. Ltd.
80 Robinson Road #02-00
Singapore 068898
Singapore
Email: subscriptions@benthamscience.net

BENTHAM SCIENCE

CONTENTS

FOREWORD

The authors have written this book based on their research in computational fluid mechanics (CFD), specifically on transient fluids. The mathematical formulation of the physical phenomena occurring in fluid pipes, the use of the method of characteristics (MOC) as a tool to solve the governing equations of the numerical models and the coding strategy of these numerical models into a CFD tool are three difficult but essential steps for any fluid mechanics' research study. These steps are well presented and detailed in this book.

The book is written in simple English and the Mathematical and Numerical formulations of the physical problems are reasonably explained. The numerical code algorithms are detailed in appendices. In brief, the book provides a simple one-dimensional method to predict what happens exactly in a fulfilled straight pipe when water hammer and cavitation phenomena occur therein. Finally, this book will surely help readers to develop their modelling skills in fluid mechanics.

Hab. Eng. Ridha ENNETTA
Mechanical Engineering
University of Gabes
Tunisia
Africa

PREFACE

Hydraulic transients also referred to as fluid transients or water hammer usually cause leakage and rupture of pipelines. Since pressure rises are responsible of these accidents, engineers, and researchers are always trying to predict the pressure history in given hydraulic plants. This purpose requires in fact accurate modelling and calculation. Perhaps, readers who are interested in fluid mechanics have met at least one complicated situation revealing water hammer. Indeed, this phenomenon is not simply solved; partial differential equations are present, where the continuity equation and the equation of motion are used. Moreover, water hammer results in transient cavitation (vaporous cavitation and gaseous cavitation), which is a combination of thermodynamic and hydraulic phenomena. Transient cavitation is destructive for hydraulic plants. The entire content of this book is a mathematical and numerical study of transient cavitation in pipelines. The emphasis is made on coupled modelling and its numerical calculation. During several years of research, the author has focused on the accuracy of the numerical result. Different models, methods and assumptions are tested in order to be close to the experimental result. These are detailed in the present book. The simulation of vaporous cavitation and gaseous cavitation by use of coupled modelling is attempted to be useful. The improvement of the numerical simulation leads to the improvement of the design of pipes in hydraulic engineering. Two types of pipelines will be considered: the quasi-rigid elastic pipelines (metal and concrete) and the VE pipelines (polyethylene). Experimental results from the literature is used to validate the proposed models. The first chapter which is a review of the literature describes the main previous works on fluid transient in pipelines. The mathematical formulations of water hammer and cavitation are detailed in the second chapter. The mathematical modelling detailed concerns water hammer and cavitation. The numerical resolution of the various models using the MOC is presented in the third and the fourth chapters. Chapter 3 describes the numerical resolution proposed for the water hammer models, while the numerical modelling of transient cavitation is the subject of the fourth chapter. The results of the numerical simulation of water hammer and cavitation in elastic pipelines are introduced and discussed in the fifth chapter where the WSA scheme is highlighted. The final chapter is reserved for the simulation of water hammer and cavitation in VE pipeline.

Abdelaziz Ghodhbani
National Engineering School of Sfax
University of Sfax
Tunisia, Africa

Ezzeddine Haj Taïeb
National Engineering School of sfax
University of Sfax
Tunisia, Africa

Mohsen Akrout
National Engineering School of sfax
University of Sfax
Tunisia, Africa

&

Sami Elaoud
National Engineering School of sfax
University of Sfax
Tunisia, Africa

INTRODUCTION

Hydraulic networks, such as water supply systems, irrigation networks, hydropower stations, nuclear power plants, petroleum plants and cooling systems in thermal power plants are usually affected by strong pressure variation due to hydraulic transients. This is a spectacular form of unsteady flow in liquid-filled pipe systems usually referred to as "water hammer". This phenomenon is mostly due to the sudden closing and opening of valves, starting, and stopping (or failure) pumps and turbines. The kinematic energy of the liquid transforms to a pressure energy following the accident. The large transient pressure variations resulting from the water hammer may damage the pipe systems and their components (junctions, valves, elbows, pumps, *etc*.). Water hammer can also lead to severe low pressures usually resulting in distributed cavitation or localized column separation. The condensation of vapour in case of vaporous cavitation or the dissolving of gas in case of gaseous cavitation involves strong collapses and subsequently hight pressure surges. Noting that the concept of "water hammer" refers to "transient without cavitation" while "transient cavitation" is a special case of fluid transient.

Cavitation is known in hydraulic engineering as the most dangerous accident for pumps, turbines, and pipelines. Prevention should be taken at critical points to avoid damage and leakage. The simplest is to minimize the flow velocity by making a wise choice of diameter. Also, prevention can be done by extending the closing times of automated or motorized valves by means of springs, dampers or magnetic brakes and the stopping time of the pumps using the flywheels. The use of anti-ram chambers with compressed air mattresses can be as effective. Another technique is to add an equilibrium chimney to the pipes of high calibre supply.

The location of critical points as well as the choice and the dimensioning of protective components require preliminary study and rigorous numerical simulation. To achieve this, mathematical modelling is necessary. All physical and dynamic parameters affecting the water hammer are taken into account: densities, modulus of compressibility of the fluid, modulus of elasticity of the pipe, initial flow velocity, initial pressure, *etc*. The mechanical behaviour of the pipe, whether elastic quasi-rigid (metals and concretes) or viscoelastic (like PVC), is also considered. The steady state defines the initial conditions of the problem whereas the physical boundary conditions of the hydraulic system are introduced into the model. In addition, fluid-structure interaction (FSI) should be considered because of the existence of three dynamic coupling modes namely: (i) fluid-wall friction (friction coupling), (ii) pipe expansion and contraction (Poisson coupling) and the inertia force of the pipe and components (junction coupling). If FSI is ignored, the calculations provide only two fluid variables namely the pressure "p" (or piezometric head "H") and the velocity "V" (or discharge "Q"). If FSI is considered, then several structural responses can be obtained besides the above fluid characteristics, such as the displacement, the pipe velocity, stresses, and strains.

There are two mathematical considerations to describe hydraulic piping systems under various hydraulic conditions [1]. The first one concerns the elastic models or water hammer models that consider fluid compressibility and pipe-wall mechanical characteristics. The pipeline is assumed to be elastic (or quasi-rigid). The elastic models are used to describe water hammer, in other word, fast changes in flow conditions, such as valve manoeuvres, pumps start-up or shutdown, or pipe bursts. For these models, the flow is usually considered one-dimensional and described by a system of differential equations. The second one deals with the rigid models or lumped models, which are valid only when the flow conditions vary slowly in time. For these models, the fluid and pipe compressibility effects can be neglected, and the fluid can be described by a rigid column. The system of hyperbolic differential equations, typical of elastic models, is simplified to a set of ordinary differential equations that can be solved by Modified Euler Method or Runge-Kutta Methods. The rigid models can be efficient for slow variations of flow, such as the increase of consumption in peak hours, in which it is important to account for the inertia of the system. The transition between the unsteady compressible flow (elastic model) and unsteady incompressible flow (rigid model) is set by the ratio between the total internal energy change and the total kinetic energy change [1].

There are various FSI models for water hammer in the literature, but the four-equation model is the most suitable among them because of its simplicity and reliability. Only axial vibration of the pipe is considered in this model; the circumferential and radial motion are ignored in much research.

Regarding viscoelastic (VE) modelling, numerous research focuses are taken on this field. The VE behaviour is usually studied by considering mechanical element to describe it; the spring for elastic response and the dashpot for viscous response. Usually, the generalized Kelvin-Voight model is used to describe VE behaviour. Fluid transient in VE pipelines, such as polyethylene can be predicted thanks to the classical VE model, in which the retarded circumferential strain is calculated. Although the important effect of the pipe material on the pressure history during fluid transient, FSI models (four-equation models) are rarely used for water hammer prediction in VE pipes. Mostly, unsteady friction (UF) is ignored because friction damping VE can be neglected against VE damping.

The cavitation models obviously differ from water hammer models because of the existence of the two-phase liquid-vapour flow. The void fraction of vapour in the mixture is introduced in the governing equations. Cavitation models are very numerous, but they can be grouped into two large families: the column separation models, for example the discrete vapour cavity model (MCVD) and the discrete gas cavity model, and the distributed cavitation models. Regardless the model type, dynamic coupling can be included in the governing equations, but it can be observed that FSI is rarely introduced in cavitation models in the literature [2]. Nevertheless, the prediction of structural responses accompanying fluid transients with cavitation requires coupled analysis in which FSI should be accurately described, especially in case of axial freely moving pipes. In practice, it is impossible to carry out hydraulic plants with entirely rigid pipelines.

In order to obtain accurate results, water hammer calculations in pipelines require the use of appropriate numerical method. The method of characteristics (MOC) is the most preferred one because the wave-speeds are constant in time (no dispersion). The majority of the commercially available water hammer software packages use the MOC. The numerical schemes are usually based on linear space or linear time interpolations. Nevertheless, since interpolation results in errors, researchers use the wave-speed adjustment scheme (WSA), which is more flexible and more efficient. The Finite Element Method (FEM) provides great flexibility in handling variable-size elements in different properties. The method consists firstly in substituting a shape function into the differential equations. Then the residual (difference between the shape function and the exact solution) is multiplied by a weighting function to obtain a weighted residual. After that, the FEM attempts to force this weighted residual to zero in an average sense. The choice of the weighting functions can be obtained with different schemes, namely the Galerkin scheme and the Petrov-Galerkin scheme. Another way is to combine two numerical methods: The FEM and the MOC. The new method is called FEM-MOC. The former is suitable for structural calculation (pipe), whereas the latter is preferred for the fluid calculation. However, this technique is not preferred than the full MOC. The Finite Difference Method (FDM) is efficient in calculation of distributed cavitation models because the pressure wave-speed is no longer constant. For implicit schemes, the FDM has the advantage of stability for large time step. However, the time and the storage requirement are increaded. Therefore, implicit schemes are not suitable for wave propagation problems because they entirely distort the path of propagation of information. Recently, hyperbolic problems are usually solved by explicit schemes, such as MacCormak and Lax-Wendroff schemes. The Finite Volume Method (FVM) is also used for hyperbolic problems, such as gas dynamic, shallow water waves and water hammer flows. The first-order based Godunov scheme is widely used for this method. It is very similar to the MOC with linear space-line interpolation.

From the foregoing description, some recommendations can be established for calculation of hydraulic transient. First, one should select appropriate model and appropriate numerical method. Secondly, some assumptions are needed for easier calculations and obviously there are some limitations for such assumptions. Thirdly, the dynamic coupling mechanisms complicate the models, and it becomes unnecessary to use them unless they lead to eventual improvement of the solution, or when structural responses prediction is needed. Finally, the computer code should be efficient otherwise, any numerical error involved cannot be easily avoided. Recently, several computer software are developed for computational fluid dynamics (CFD), and fluid transient problems represent the most important application for them. The most popular CFD products are Matlab and Ansys Fluent. The majority of reviewers stated that Matlab is designed for numerical computing and visualization but is not suited for intensive computation. However, Ansys Fluent is essentially directed for modelling and simulation thanks to its efficient and automated structured meshing. In this work, numerical computations of numerous variables is needed, all calculations are hence performed by use of Matlab through implementation of several codes written by the present authors.

In this work, a coupled modelling and numerical calculation of fluid transient in pipelines is proposed. The reservoir-pipe-valve system (downstream valve system) is mainly used in defining boundary conditions of the problems. The emphasis is made on cavitation (vaporous cavitation and gaseous cavitation), its effect on the pressure history and especially on the structural responses. Water hammer and cavitation models from the literature will be used in new coupled formulation that considers the effect of FSI on the numerical solution. Friction coupling effect will be tested by comparison between steady friction (SF) and UF. Poisson coupling will be investigated by use of the four equation model. To introduce the junction coupling, axial free-moving valve will be considered and compared to the fixed valve case. Elbows systems lead to higher junction coupling and additional lateral vibration of the pipe. Such configurations are more complicated and may be investigated in next works. The objective is to improve water hammer understanding associated with local column separation.

NOMENCLATURE

Abbreviations

4EM	Four-equation Model
4EFM	Four-equation Friction Model
4EVEM	Four-equation Viscoelastic Model
BPA	Benchmark Problem A
CCC	Constant creep-compliance
CWS	Constant wave-speed
CFD	Computational Fluid Dynamics
DVCM	Discrete Vapour Cavity Model
DGCM	Discrete Gas Cavity Model
FDM	Finite Difference Method
FEM	Finite Element Method
HDPE	Hight Density Polyethylene
HGL	Hydraulic Grade Line
MOC	Method of Characteristics
QR	Quasi-rigid
RG	Rectangular Grid
SF	Steady friction
SG	Staggered Grid
SLI	Space-line Interpolation
TLI	Time-line Interpolation
UF	Unsteady friction
CCC	Constant creep-compliance
VCC	Variable creep-compliance

VWS	Variable wave-speed
VEM	Viscoelastic Model
VE-DVCM	Viscoelastic Discrete Vapour Cavity Model
VE-DGCM	Viscoelastic Discrete Gas Cavity Model
WSA	Wave-speed Adjustment

Scalars

α	Vapour void fraction / constant
$\forall$	Volume
β	Constant number / coefficient of the Newmark method
δ	Integer equal to 1 or 2
κ	Isothermal compressibility
ρ	Mass density
ψ	Weighing factor
Ω	Number
ε	Axial strain
γ	Pipe inclination / constant
λ	Characteristic direction
μ	Dynamic viscosity
ν	Poisson coefficient / kinetic viscosity
ψ	Weighting factor
σ	Stress
Σ	Constant number
Π	Constant number
Γ	Constant number

τ	Shear stress / retarded time
Φ	Number
Φ	Airy function
ζ	Damping ratio
a, b	Integers
A, $\tilde{A}$, $\hat{A}$	Cross section
B	Pipeline impedence
c	Anchor coefficient / viscous damping coefficient
C	Celerities / constant number
D	Inner diameter of the pipe
e	Pipe-wall thikness
E	Young's modulus of elasticity
f	Friction coefficient
F	Force / number
G	Number
g	Gravity acceleration
h	Friction head loss / arbitrairy datum
H	Piezometric head
i	Space increment
j	Time increment
J	Creep compliance
k	Constant / number / stiffness coefficient
K	Bulk modulus of the liquid / number
L	Length of the pipe
m	Mass

n	Rational number
N	Number of reaches
p	Pressure
P	Computational point
$q, \tilde{q}, \hat{q}$	discharge
Q	One time-step earlier computational point
R	Radius / two-step earlier computational point
r, R	Radius
t	Time
T	Temperature / period
u	Axial displacement / integration time
V	Celerity
v	Specific volume
W	Weighting function
$Z, \tilde{Z}, \hat{Z}$	Head
z	Axial cordinate

Matrices

A, B	Coefficient matrices of the PDE system
D	Diagonal matrix
I, $\bar{\bar{I}}$	Identity Matrix
M1, M2, M3, M4	Coefficient matrices of the algebraic system for inner sections
R1, R2, R3, R4	Coefficient matrix of the algebraic system at the reservoir
T, S	Transformation matrices

V1, V2, V3, V4	Coefficient matrix of the algebraic system at the valve

Tensors

$\overline{\overline{D}}$	Strain tensor
$\overline{\overline{\tau}}$	Viscosity stress tensor
$\overline{\overline{\sigma}}$	Cauchy's stress tensor
$\overline{\overline{\overline{\eta}}}$	Orientation tensor

Vectors

k1, k2, k3, k4	Right-hand side vectors of the algebraic system for inner sections
r1, r2, r3, r4	Right-hand side vectors of the algebraic system at the reservoir
v1, v2, v3, v4	Right-hand side vectors of the algebraic system at the reservoir
r	Right-hand side vector of the PDE system
y1	Vector of unknowns of the water hammer models
y2	Vector of unknowns of the column separation models

Subscripts

0	Initial
b	Beat
cav	Cavitation
d	Downstream
e	Elastic

f	Fluid
g	Gas
k	Kelvin-Voight element
l	Liquid
m	Mixture
p	Pipe
v	Vapour
r	Radial coordinate / retarded
s	Steady-state / integration time
T	Isothermal
u	Unsteady-state / upstream
w	Pipe-wall
z	Axial coordinae
φ	Circumferential coordinate

CHAPTER 1

Review of Literature

Abstract: This chapter presents a review of the literature related to fluid transients involving column separation and vaporous and gaseous cavitation in pipelines. Basic concepts are clarified, and the classification of water hammer and cavitation models is presented. The historical development of fluid transients is described. The concept of propagation of pressure pulses at sonic velocity through a straight pipeline is presented.

Keywords: Cavitation; Column separation; Experiment; Water hammer.

INTRODUCTION

Fluid transients in pipelines are an important topic in hydraulic engineering that needs deep studies and research to prevent fluid and pipe responses. Numerous researchers define mathematical and numerical models for this purpose. Others tried to obtain the real behaviour through experiments. In fact, these experiment results serve as references to compare numerical results and validate mathematical models and numerical methods. The aim of this chapter is to focus on the general concepts of fluid transient studies.

FLUID TRANSIENTS

Fluid transients is a large variation in pressure and velocity following an event in the hydraulic system stopping or starting the fluid flow. It is also referred to as water hammer in pipelines. Generally, the phenomenon occurs in full-liquid pipelines and is caused by sudden valve closing, pump stopping and starting, *etc*. As an outcome, fluid transients may lead to severe accident for the hydraulic plants. The concept of fluid transients began in the middle of the nineteenth century by Ménabréa [3, 4] and has continued after the middle of the twentieth century.

Water Hammer Pressure Rise

The water hammer law in a straight pipeline was established by Joukowsky in 1900. The increment of pressure change Δp depends on the increment of flow velocity ΔV *via* [5].

Abdelaziz Ghodhbani, Ezzeddine Haj Taïeb, Mohsen Akrout & Sami Elaoud

$$\Delta p = -\rho_f C_f \Delta V \quad \textbf{(1.1)}$$

with ρ_f is the mass density of the fluid and C_f is the pressure wave-speed. Since $\Delta p = \rho_f g \Delta H$, in which ΔH is the head change, Eq. (1.1) implies [6].

$$\Delta H = -\frac{C_f \Delta V}{g} \quad \textbf{(1.2)}$$

Elastic waves propagate without modification in an infinite isotropic medium but are susceptible to reflection and refraction when they meet a surface separating two different media. Two cases are to consider: (i) When a plane wave propagating in a fluid, normally meets a rigid surface, it is reflected without changing sign: thus, a compression wave is reflected in a compression wave. (ii) When a plane wave propagating in a fluid, normally meets a surface where the pressure remains constant (free surface of a liquid), there is reflection with change of sign: thus, a wave of compression is reflected in a rarefaction wave.

Water Hammer Process

Consider the case of the standard reservoir-pipe-valve-system containing a fluid (*e.g.* water) flowing at the initial velocity V_0 from the reservoir up to the valve. The sudden closure of the valve causes water hammer. By neglecting the friction throughout the pipe, the propagation of water hammer is subdivided into four events (Fig. **1.1**) as follows:

Event 1: At the instant $t = 0$, the valve at the downstream end of the installation is closed. Immediately and close to the valve, the flow rate vanishes. The fluid nearest the valve is compressed to the extra head H and the pipe-wall is stretched (Fig. **1.1a**). As soon as the first fluid layer is compressed, the process is repeated for the next fluid layer. The fluid upstream from the valve continues to move downstream until the compression of the whole fluid column. From an energetic point of view, the kinetic energy of the stopped fluid is transformed into elastic energy, part of which serves to deform the pipe (expansion) and the other to compress the fluid. When the wave reaches the reservoir, at $t = L/C_f$ the pressure is uniform, and all the fluid is under the extra head H and all the kinetic energy is transformed into elastic energy.

Event 2: To restore the equilibrium between the pressures of the reservoir and the pipe, the fluid flows backward at the same initial velocity since the pressure of the reservoir is constant. A low-pressure wave is superimposed to the existing overpressure (Fig. **1.1b**). At $t = 2L/C_f$, the kinetic energy of the previous event stored in the form of elastic strain energy is restored. The pressure is back to normal along the pipe (to H_0) and the velocity is equal to V_0.

Event 3: The fluid continues to flow towards the reservoir and subsequently the pressure decreases at the valve by H meters lesser than H_0. The low-pressure wave reflected by the closed valve travels back to the reservoir (Fig. **1.1c**). The pressure is set to $H_0 - H$ on the entire pipe at $t = 3L/C_f$. Then, the fluid is expanding while the pipe is contracting in latera direction.

Event 4: The unbalanced conditions at the reservoir cause the fluid to flow downstream at the velocity V_0 and a positive-pressure wave progresses forward to the valve. The pipe and the fluid return to normal conditions (Fig. **1.1d**). At the instant, the high-pressure wave reaches the valve, the fluid and pipe conditions are the same as $4L/C_f$ seconds earlier. This process is then repeated every period $T = 4L/C_f$.

Water hammer process can be visualized in the space-time plane as illustrated in Fig. (**1.2**). The sloped dashed lines represent the wave fronts corresponding to either high-pressure or low-pressure waves with increasing time. These lines divide the space-time plane into zones at different pressures and different velocities.

Noting that the analysis of water hammer process in case of a sudden stoppage of a pump, of a flow in a discharge pipe is the same except, however that it begins with a pressure drop and ends with an overpressure. It would be enough to repeat the explanations given above starting with event 3.

Pressure wave-speed

The pressure wave-speed depends on the physical properties of the fluid and the mechanical and geometrical characteristics of the pipe. Its determination requires the search for the celerity of the acoustic wave in the fluid in question. The starting formula concerns the blood flow in the vessels as $C_{f0} = \left(K/\rho_f\right)^{1/2}$ with K is the bulk modulus of the fluid.

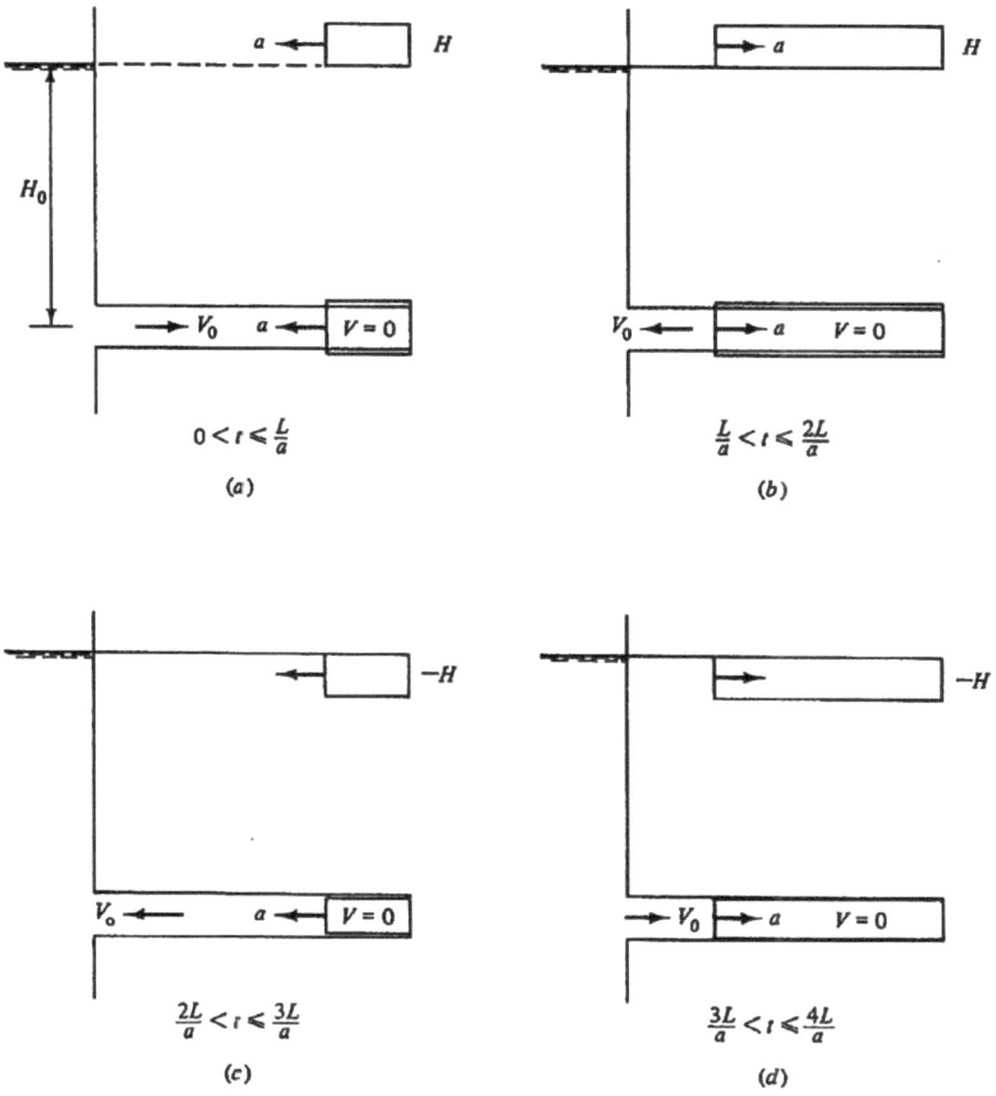

Fig. (1.1). Sequence of events for one period after sudden closure of a valve [6].

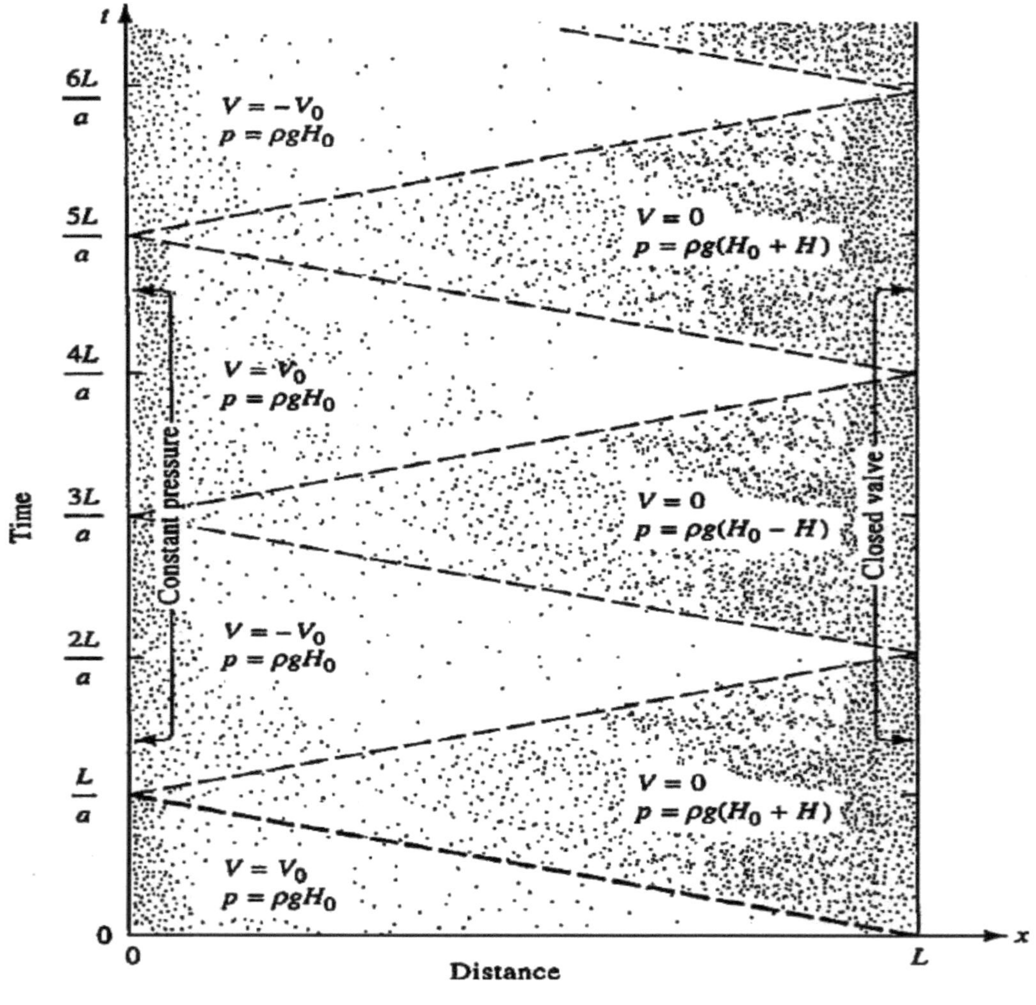

Fig. (1.2). Fluid transient in the space-time plane [6].

The pipe-wall elasticity represented by the Young's modulus is in fact constant. For incompressible fluids in elastic pipes, Weber and Moens established that $C_{f1} = \left(Ee/\left(\rho_f D\right)\right)^{1/2}$ with E denotes the Young's modulus of elasticity, e is the pipe-wall thickness and D is the inner diameter of the pipe [7 and 8]. Korteweg gave the general expression of the pressure wave-speed for compressible fluids in elastic pipes as [9].

$$\frac{1}{C_f^2} = \frac{\partial \rho_f}{\partial p} + \frac{\rho_f}{A_f}\frac{\partial A_f}{\partial p} \tag{1.3}$$

with A_f is the cross section. For rigid pipes, the pressure wave-speed is obtained by neglecting the term $\partial A_f/\partial p$ in Eq. (1.3). In addition, for incompressible fluids flowing in elastic pipes, the term $\partial \rho_f/\partial p$ can be neglected and the fluid property is defined by [9].

$$\frac{\partial \rho_f}{\partial p} = \frac{\rho_f}{K} \tag{1.4}$$

The theory of elasticity was used to express the term $\partial \rho_f/\partial p$ by neglecting the axial stress in the wall, the axial displacement as well as the inertia effect of the pipe [9]. This last hypothesis is validated for rigidly supported pipes with seals. A static equilibrium relationship between the forces acting on an axial element of the pipe leads to $Ddp = 2ed\sigma_\varphi$, with σ_φ denotes the circumferential stress. If the radial stress σ_r is ignored, elasticity laws lead to $Ed\varepsilon_r = d\sigma_\varphi$, with ε_r is the radial strain. The expressions of A_f and ε_r allow the differential equation $2dA = \pi d\varepsilon_r D^2$, which can be combined with the two previous relations to obtain the following:

$$\frac{\partial A_f}{\partial p} = \frac{\pi D^3}{4Ee} \tag{1.5}$$

The celerities verify $1/C_f^2 = 1/C_{f0}^2 + 1/C_{f1}^2$ and subsequently,

$$C_f = \left[\rho_f \left(\frac{1}{K} + \frac{D}{Ee}\right)\right]^{-1/2} \tag{1.6}$$

Parmakian established that axial stress and axial strain of the pipe have a significant effect on the pressure wave-speed C_f [10]. He showed that C_f depends on the Poisson's ratio and the restriction imposed on the axial deformation of the pipe. For the same reason, Streeter and Wylie presented a detailed development of C_f by considering the type of pipe support and the movement of the pipe [11]. For this, they cited three anchoring conditions for the axial deformation of the pipe, which are frequently used to approach several practical situations. A correction is made to the formula (1.6) to give the new expression of C_f as:

$$C_f = \left[\rho_f \left(\frac{1}{K} + \frac{c_1 D}{Ee}\right)\right]^{-1/2} \tag{1.7}$$

with c_1 is the anchor coefficient (see subsection 2).

FLUID-STRUCTURE INTERACTION

Korteweg considered that pressure induces circumferential stress and radial displacement in the pipe [9]. However, only Poisson effect is considered when axial inertia of the pipe is not neglected; axial stress wave propagates along the pipe. Gromeka introduced the pipe inertia by considering incompressible fluid in elastic pipe [12]. He defined two waves: a pressure wave in the fluid and axial stress wave in the pipe. Lamb (1898) mentioned that the pipe and the fluid interact during water hammer [13]. The axial stress wave- speed is defined by

$$C_p = \left(\frac{E}{\rho_p}\right)^{1/2} \tag{1.8}$$

with ρ_p is the density of the pipe-wall material and E is the Young's modulus of elasticity.

In 1956, Skalak followed the Lamb's studies and provided an excellent work used as reference for the FSI analysis [14]. He defined two waves: the pressure wave in the fluid and the axial stress wave in the pipe. Skalak's theory demonstrated the existence of a precursor wave that precedes the arrival of the water hammer wave. In 1969, Thorley was the first who observed the precursor wave following his experiment on steel, aluminium, and polyethylene pipes of 14 m long, 0.05 m in diameter and 0.005 m thick (Fig. **1.3**) [15].

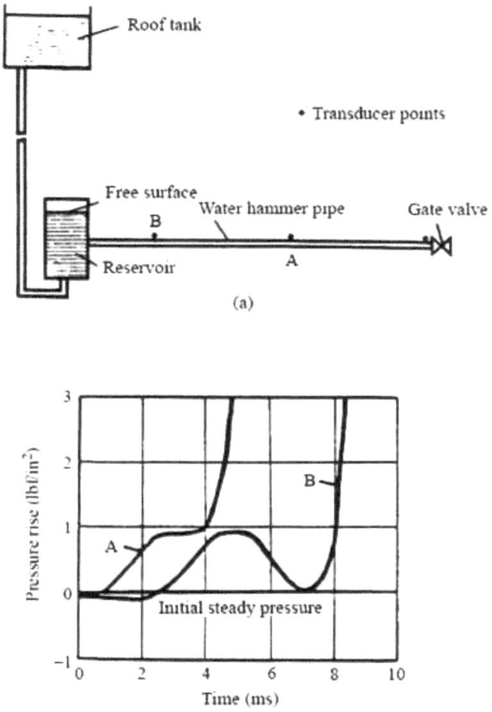

Fig. (1.3). Schematic diagram of water hammer apparatus (**a**). Precursor waves in two different points A and B [16].

The classical theory of water hammer ignores the effect of several structural parameters on the fluid. FSI is only manifested by the dependence of the pressure wave-speed on the anchoring conditions of the pipes defined by the coefficient c_1. Such a parameter is expressed with respect to the unique variable v (Poisson's ratio). The anchoring conditions of the pipe impose the axial displacement u_z (or strain ε_z) as well as the axial stress σ_z. In theoretical analysis, three support conditions are standard [16]:

i) Pipe anchored with expanded joints throughout its length: $\sigma_z = 0$, $\varepsilon_z = 0$ and $c_1 = 1$

ii) Pipe anchored throughout axial motion: $\sigma_z \neq 0$, $\varepsilon_z = 0$ and $c_1 = 1 - \nu^2$

iii) Pipe anchored at its upstream only: $\sigma_z \neq 0$, $\varepsilon_z \neq 0$ and $c_1 = 1 - \nu/2$

The classical theory of water hammer is valid only for the first two cases which do not present axial motion. The modern water hammer theory expresses FSI by dynamic interaction between four waves [17]: pressure wave in the fluid and three stress waves in the pipe-wall. The interaction between all these waves takes place at junction points (valves, elbows, tee, *etc.*). This mechanism is referred to as *junction coupling*. Throughout the pipe, the interaction is established between the pressure wave and the axial stress wave. This is called *Poisson coupling* since it is due to the radial contraction-expansion mechanism of the pipe. The third mechanism of FSI is due to the relative internal motion between the pipe-wall and the fluid, it is called *friction coupling*.

Fig. (**1.4**) shows a fluid in normal flow in a pipe at given initial velocity and given initial pressure. The sudden closing of the valve causes a pressure rise in the fluid. The local radial expansion of the pipe-wall generates radial contraction propagating throughout the pipe. As a result, a second pressure wave of low magnitude travels at the axial stress wave-speed C_p (referred to as c_t in Fig. (**1.4**). This corresponds exactly to the precursor wave already defined in [14, 15].

Junction coupling phenomenon is created at specific points (valves, elbows, tee, *etc.*). The radial contraction-expansion mechanism of the pipe caused by the pressure wave induces in the fluid a new pressure wave which in turn affects the pipe motion and so on. Fig. (**1.5**) illustrates a standard system called vibrating elbow. Fig. (**1.5a**) shows the steady state regime. When a pressure wave arrives at the right elbow (for example), the pressure rise causes the system to stretch to the right (Fig. **1.5b**), while the pressure wave continues to propagate to the left at the pressure wave-speed C_f . This axial motion induces a second pressure rise at the left elbow. The new pressure wave propagating to the right at the same wave-speed C_f causes a second motion of the pipe to the left, and the cycle is repeated. The pressure magnitude wave gradually decreases until the system stabilizes.

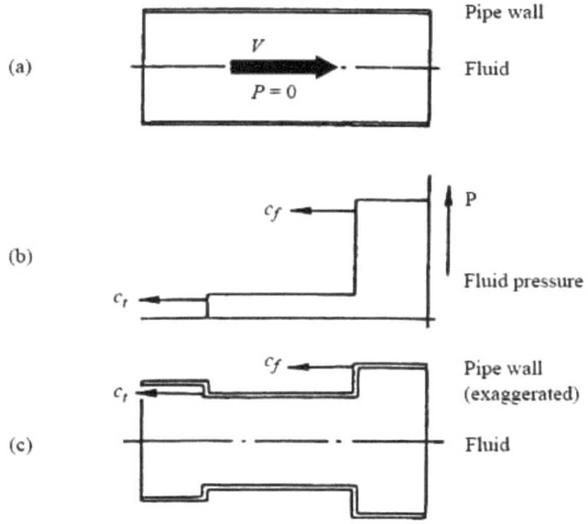

Fig. (1.4). Poisson coupling mechanism [16].

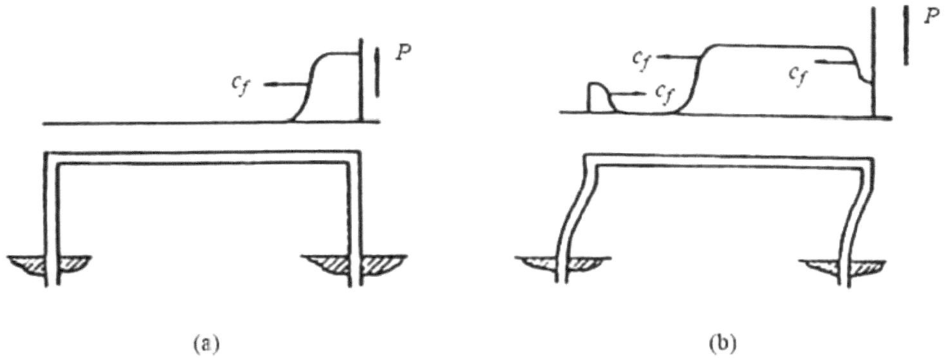

Fig. (1.5). Junction coupling mechanism for vibrating one-elbow system [16].

TRANSIENT CAVITATION

Cavitation in liquids is a broad field of research. This subsection is confined to *transient cavitation* also referred to as *water hammer induced cavitation*. According to the occurrence of gas release phenomenon, cavitation can be either vaporous or gaseous. In addition, the manner of cavity forming is important. If cavity occurs at localized sections of the pipeline, the phenomenon is called *column separation*. In contrast, if bubbles are distributed throughout the pipe, then *distributed cavitation* is considered. Regardless the type of cavitation, the high-pressure spikes induced can cause damage of the pipeline system and its components.

Vaporous Cavitation

Vaporous cavitation supposes the forming of vapour when the pressure drops below the vapour pressure of the liquid. For instance, the state diagram of water Fig. (**1.6**) shows occurrence of vaporous cavitation resulting from pressure drops in the liquid. Two types of vaporous cavitation can be observed, the void fraction α can be used to identify the type of cavitation. Wallis (1969) defined the void fraction α as the ratio of the vapour volume $\forall_v$ to the total volume of the vapour-liquid mixture $\forall$ [18].

$$\alpha = \frac{\forall_v}{\forall} \tag{1.9}$$

The void fraction depends on the magnitude of the velocity gradient in the cavitating flow. According to the value of α, two types of vaporous cavitation can be distinguished: (i) the discrete vapour cavity or *local liquid column separation* and (ii) the distributed vaporous cavitation or *bubbly flow*. The former corresponds to large values of α whereas the latter occurs for the small values.

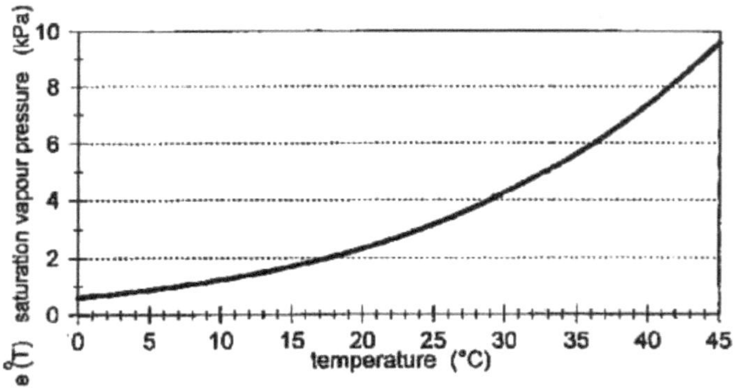

Fig. (1.6). Pressure and temperature of water at saturation.

Gaseous Cavitation

The concept of gaseous cavitation has been started in the 1970s and early 1980s by introducing the effect of dissolved gas and gas release in transient studies [19, 20, 21]. The liquids have the property of dissolving some amount of a gas meeting them through a free surface. The gas comes out of solution when the pressure drops below the saturation pressure. Gas release occurs in several types of hydraulic systems, such as cooling water systems, long pipelines with high points, oil pipelines, *etc.* [22]. The time of gas release takes several seconds and is longer than the time of vaporous cavity forming, which takes only few milliseconds. However, gas absorption is slower than gas release (order of minutes) [23]. It is assumed that released gas stays in the cavity and does not immediately dissolve following pressure rises. The presence of entrained air or free gas in the liquid reduces the pressure wave propagation.

As described for the vaporous cavitation, the void fraction of released gas can be also defined for gaseous cavitation as the ratio of the gas volume $\forall_g$ to the volume of mixture $\forall$ [6].

Unlike the monophasic flow, in which the pressure wave-speed is assumed to be constant, the pressure wave propagation in gas-liquid mixture depends on the absolute partial pressure p_g^* of the gas, and it is generally significantly lower than the liquid pressure wave-speed C_f . By considering some approximations, the pressure wave-speed $C_{f.m}$ in the gas-liquid mix with respect to C_f is defined as [6].

$$C_{f.m} = C_f \left(1 + \alpha \frac{\rho_l C_f^2}{p_g^*} \right)^{-1/2} \tag{1.10}$$

However, this formula makes the hyperbolic system highly nonlinear and difficult to solve with the MOC schemes. Alternatively, the distributed free gas can be lumped to the computational sections leading to the DGCM.

Liquid Column Separation

Liquid column separation is usually accompanying water hammer in liquid filled pipelines. When the pressure drops under the vapour pressure at specific locations due to the propagation of the low-pressure wave, liquid column separation takes place with a high celerity gradient of the adjacent liquid. Vapour cavities (for vaporous cavitation) or gas cavities (for gaseous cavitation) coalesce and form a unique cavitation zone where the void fraction is close to unity. The cavity length increases by inertia, then it is stopped by the low pressure. When the low-pressure wave arrives to the reservoir and is reflected to a high-pressure wave, the cavity is compressed, and vapour condenses therein. This phenomenon is referred to as *cavity collapse* which induces *shock waves*. The local pressure usually reaches high peaks. In his thesis work presented in 1986, Simpson studied the evolution of short duration pressure pulse due to cavity collapse [24]. These high-pressure peaks can lead to severe damage in the hydraulic system. Surge tanks, air chambers or large motor rotating of inertia can be used to avoid liquid column separation.

Column separation can occur in two modes: (i) local cavity and (ii) intermediate cavity. Local cavities were studied by Joukowsky in 1900 who gave experimental results for them (Fig. **1.7**) [5]. They take place at specific locations, such as valves, closed ends, pumps, elbows, *etc*.

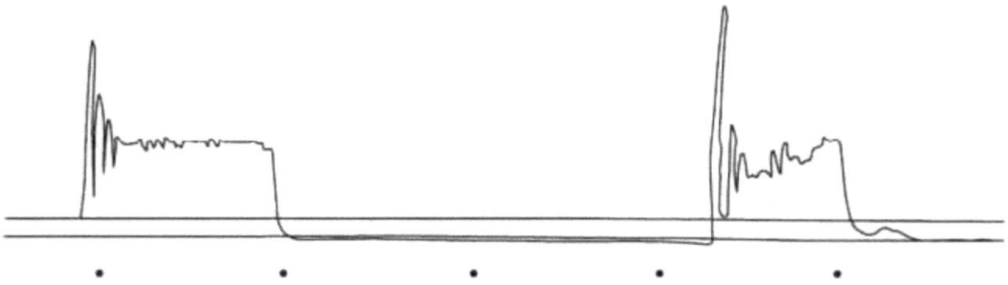

Fig. (1.7.) Local vapour proposed by Joukowsky (1900). Horizontal top line: steady state pressure, Horizontal bottom line: barometric pressure [25].

Hogg and Trail [26] and Langevin [27] followed the Joukowsky's works on local column separation. LeConte presented an experimental study and analytical development on this phenomenonin 1937 [28]. Intermediate cavity occurs at critical points following the superimposing of two low-pressure waves of opposite directions [29]. In 1959, in a thesis work at the Melbourne university, O'Neill described the intermediate vapour cavity using a graphic method [30]. He pointed out that this type of column separation is an internal boundary condition in the pipe. Pressure rises induced by cavity collapses were observed. Since 1960 up to 1965, Sharp followed the O'Neill's works by considering ideal spherical cavity [31, 32 and 33]. Jordan (1961) demonstrated the occurrence of intermediate vapour cavities following pump stopping [34]. He developed an analytical method to localize vaporous cavities accurately. Later, Intermediate cavity forming was visualized, and the short duration pressure pulse was observed in the experiment [24, 35, and 36].

Distributed Cavitation

Distributed cavitation is a diphasic zone of liquid-vapour mixture overlying a long portion of the pipe. It is also referred to as two-phase bubble flow. The vaporous cavitation region may result from the passage of a negative pressure wave through part of the pipeline. Decrease in the pressure change results in a smaller change in velocity while the void fraction in the cavitation zone is smaller than unity [24].

The difference between distributed cavitation and column separation is described in [37]. In 1939, Knapp developed the concept of distributed cavitation and pointed out that the graphical method cannot describe this phenomenon [37]. The occurrence of distributed cavitation is illustrated in [6] and [24]. Lupton (1953) described some events of distributed cavitation and distinguished it from column separation [29]. He established that the axial celerity gradient results in the forming of numerous liquid columns where the pressure of their interfaces is equal to the vapour pressure plus the partial pressure of dissolved gases. In 1965, Jordan studied both column separation and distributed cavitation in different sloping pipes, and he developed an analytical method for distributed cavitation description in which the pipe inclination was considered [38]. The study was experimentally validated. Distributed cavitation may result from the passage of negative wave throughout a portion of the pipe in which the pressure decreases with the wave propagation because of friction or pipe sloping [39]. In contrast, if the pressure is increasing in the same direction of the negative wave propagation, vaporous cavitation does not occur [24, 35 and 40].

The shape and inclination of the pipe determine whether cavitation is concentrated (column separation) or distributed (Fig. **1.8**). The smaller the velocity gradient is, the smaller the vapour void fraction is.

Two-Phase Flow Regime

The diphasic fluids are different from monophasic fluids regarding mass density, acoustic property, and pressure wave-speed. In this subsection, the pressure dependence of the void fraction in two-phase flow regime is described.

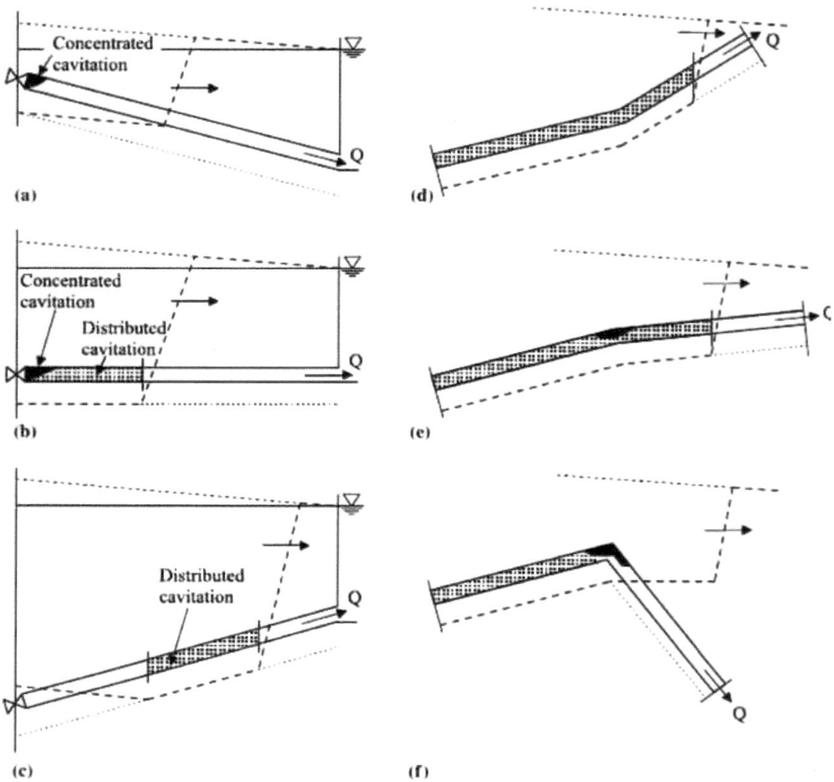

..... *Vapour pressure;* - - - *Steady state HGL;* - - - *Transient HGL*

Fig. (1.8). Vaporous cavitation for different geometries of pipelines [25].

In 1977, Kieffer studied the celerity of sound in two-phase flows where a mixture of saturated water and vapour are presented in the entropy diagram (Fig. **1.9**) [41]. The state G is a liquid-vapour equilibrium. The chord ratio FG/HG defines the mass fraction of vapour in the mixture. Isentropic pressure changes (compression and rarefaction) can occur during propagation of sound wave and are represented by movement up and down the constant entropy line CGK. If liquid and vapour remain in thermal equilibrium on the saturation line, then condensation and evaporation must take place in order to allow mass transfer between phases. However, Kieffer established that occurrence of these thermodynamic evolutions is not instantaneous since heat and mass transfers occur at finite speeds in 1977. The time lag in condensation or evaporation is important in determining the degree of equilibrium obtained in the sound wave.

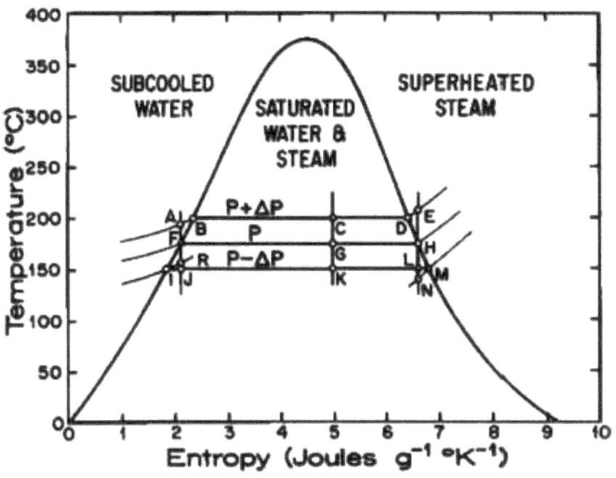

Fig. (1.9). Isentropic processes of liquid-vapour mixture in the entropy diagram of water [41].

Cavity Collapse and Short Duration Pressure Pulse

Cavity collapse consists in the movement back of the liquid column and condensation. Shock wave and short duration pressure pulse can result from this collapse. In 1937, Angus defined this phenomenon at a closed valve [42]. The collapse process in a standard reservoir-pipe-valve system was described in a duration $t < 2L/C_f$ [24 and 43]. The pipe of length L is fully filled by a liquid flowing at the initial celerity V_0 and maintained at the constant pressure head H_0. The effect of friction was neglected in this analysis. First, the sudden closure of the valve results in the Joukowsky pressure rise (Fig. **1.10**). The compression wave propagates back towards the reservoir and then the rarefaction wave comes back to the valve at $t = 2L/C_f$ where it results in a pressure-head drop. If this pressure-head drop reaches the vapour pressure of the liquid, then the velocity in the reverse direction is no longer reduced to zero, but it decreases by ΔV_{vc} given as [43].

$$\Delta V_{vc} = \frac{g}{C_f}(H_0 - H_v) \tag{1.11}$$

with $H_v = p_v/(\rho_f g)$ denotes the vapour pressure head for horizontal pipes. Hence, the liquid flows towards the reservoir and due to the liquid column inertia, column separation takes place at the valve leading to the forming of vaporous cavity therein. The cavity acts as a fixed-pressure boundary condition and the water hammer wave continues to propagate to and from the reservoir.

Once the liquid celerity increases at the valve, the cavity begins to shrink until it finally collapses (point A in Fig. (**1.10**). Usually, the cavity collapse occurs at a non-multiple of $2L/C_f$ seconds. The rarefaction wave reflected by the cavity at time $4L/C_f$ reaches the reservoir, where its sign is inversed and propagates towards the valve. The superimposing of the compression wave to the collapse wave (shock wave) at time $6L/C_f$ leads to the short duration pressure rise (point B in Fig. (**1.10**). This pressure rise may exceed in certain cases the Joukowsky rise. Finally, upon the reflected shock wave is inversed at the reservoir and propagates upstream the valve as a rarefaction wave, the pressure decreases by the inverse of the collapse rise and reaches the same spike of point A. Since the liquid continue to flow towards the reservoir due to the reflected rarefaction wave, the pressure continues to decrease by the same pressure change pf the collapse. The superposition of this pressure drop to the pic of point B gives the final pressure stage until the time $t = 8L/C_f$.

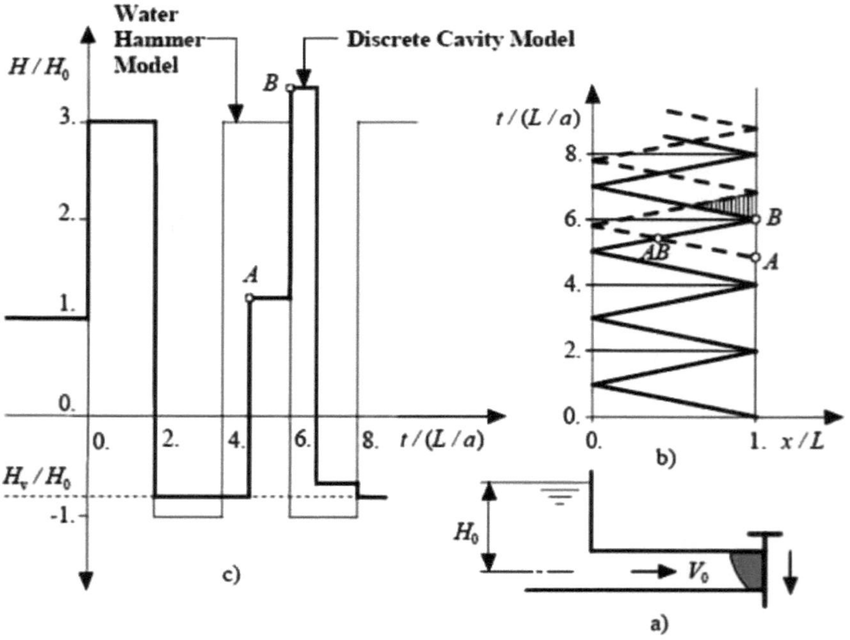

Fig. (1.10). A short duration pressure pulse. (**a**) Reservoir pipe-valve-system. (**b**) wave paths in distance-time plane. (**c**) Piezometric head history at valve [43].

FLUID TRANSIENT MODELLING

The Water Hammer Classical Model

The classical theory of water hammer dealt with the reference model. The development of the continuity equation and the momentum equation of the fluid leads to the following partial differential equations.

$$\frac{\partial H}{\partial t}+\frac{C_f^2}{g}\frac{\partial V}{\partial z}=0 \tag{1.12}$$

$$\frac{\partial H}{\partial z}+\frac{1}{g}\frac{\partial V}{\partial t}+\frac{fV|V|}{2gD}=0 \tag{1.13}$$

with D is the inner diameter of the pipe and f is the friction coefficient. The use of the method of characteristics (MOC) allows the compatibility equations

$$\pm\frac{dH}{dt}+\frac{C_f}{g}\frac{dV}{dt}+\frac{fC_f}{2gD}V|V|=0 \tag{1.14}$$

The integration of the compatibility equations along the characteristic directions is given in appendix C.

The Water Hammer Fsi Model

Despite the good results provided by the classical model, the structure behaviour (stress, velocity, displacement) cannot be predicted. Hence, the extended water hammer theory dealt with fluid and structure equations. The coupled modelling of water hammer is attributed to R. Skalak in 1956 [14]. The four-equation model (4EM) describing axial vibration of the coupled fluid-pipe system is the most popular FSI model is used as reference in FSI research. The considered pipe is of inner diameter $D = 2R$ (The radius R and the thickness e are referred to as a and h respectively in Fig. (**1.11**), a thickness e, a mass density ρ_p , a Young's modulus of elasticity E and a Poisson ratio ν (Fig. **1.11**). The pipe contains the fluid of mass density ρ_f and bulk modulus of compressibility K. On the right side of the pipe, the axial stress and the axial displacement are negligible. The pressure waves propagate uniformly in quasi-rigid elastic pipelines but not in elastic pipes. Skalak (1956) defined the four-equation model and mentioned that the solutions are pressure waves without dispersion. His model was experimentally validated by numerous researchers [44 and 45].

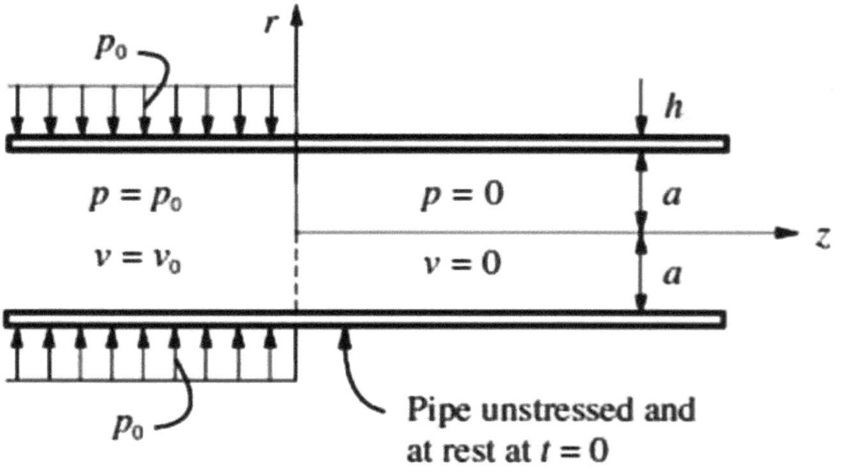

Fig. (1.11). Initial conditions of the Skalak's problem [46].

The mathematical formulation involves the continuity principle and the energy conservation principle. The water hammer theory for the fluid and the Timoshenko beam theory and elasticity theory (Hooke's law) for the pipe are combined and developed. Thus, two governing equations for the fluid are coupled to two governing equations for the pipe by means of boundary conditions representing the contact between fluid and pipe-wall on the interface. The full-MOC and the combined MOC-FEM can be used to solve the 4EM. The standard mono-dimensional problem consisting of a straight pipe with an instantaneous closing valve is considered. Four unknowns are calculated in each iteration: pressure p (or piezometric head H) and velocity V for the liquid and axial stress σ_z and axial velocity for the pipe. The governing equations of the 4EM will be detailed in the next chapter.

Some assumptions are adopted when solving the 4EM. First, the transient flow is one-dimensional (pressure p and velocity V are uniform at each section A of the pipe). Secondly, the contained liquid is assumed to be Newtonian with homogeneous, isotropic, and linearly elastic properties, and the pipe is made of an isotropic homogeneous material. In addition, pressure waves propagate at low frequencies in axial direction z, and radial inertia effect are neglected.

The 4EM was validated for acoustic behaviour of low frequencies of thin wall straight with circular section. The linear elastic behaviour was considered. Tijsseling (2003) proposed an exact solution for the model and compared the results against the experiment of Wilkinson and Curtis in 1980 where Poisson and junction coupling were highlighted [45]. The numerical results were validated using experimental measurements carried out by the researchers. In 1999, Haj Taïeb demonstrated the effectiveness of the Lax-Wendroff finite difference method (FDM) and the finite element method (FEM) in solving the 4EM applied to elastic and viscoelastic pipes [47].

The Classical DVCM

The DVCM is used for most engineering transient simulation software packages because of its simplicity and its reproducibility of many features of column separation. The DVCM allows vapour cavities to form at computational sections if the absolute pressure is computed to become below the vapor pressure p_v of the liquid (Fig. **1.12**). A constant pressure wave-speed is assumed for the liquid between computational sections. These computational sections referred to as cavities locations are treated as fixed internal boundary conditions and the pressure is set equal to the vapour pressure of the liquid until the cavity collapses. The classical DVCM can give accurate results if the number of reaches is restricted; unrealistic peaks may be generated due to multi-cavity collapses [36]. Different methods used to attenuate unrealistic pressure spikes due to multi-cavity collapses are classified in [43]. Wylie and Streeter (1993) described The DVCM is described in details and a FORTRAN computer code is given in [6]. Experimental investigation was performed by Sharp in 1977 [48]. In 1990, Tijsseling and Lavooij presented calculation and numerical results by considering the column separation with FSI in the standard reservoir-pipe valve system [49]. In 1992, Fan and Tijsseling have successfully observed the Poisson coupling induced cavitation [50]. Experimental and numerical results carried out on elbow-pipe system are presented in [16 and 51].

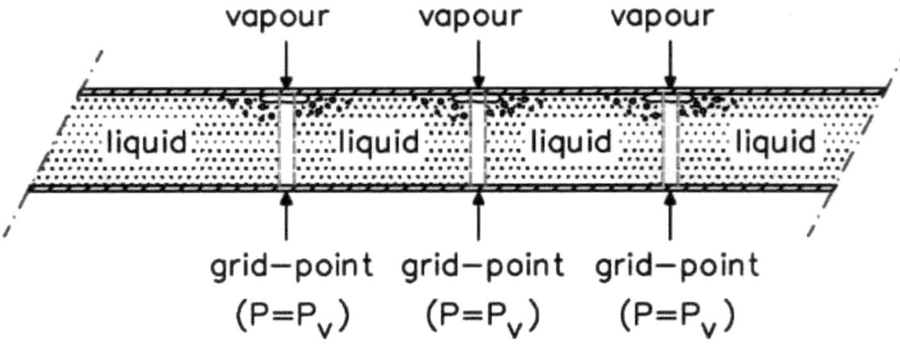

Fig. (1.12). Definition sketch of the DVCM [22].

The calculation of the DVCM is assumed to be simple since one need just a small modification on the compatibility equations Table **5.1**. In a staggered grid (Fig. **1.13**), in which P is the computation point and A and B are previous points, the compatibility equations, along characteristic directions C_f^+ and C_f^- are [52]:

$$C_f^+ : \ H^P - H^A + \frac{C_f}{gA}\left(q_u^P - q_d^A\right) = 0 \tag{1.15}$$

$$C_f^- : \ H^P - H^B - \frac{C_f}{gA}\left(q_d^P - q_u^B\right) = 0 \tag{1.16}$$

with q_u^P and q_d^P are respectively upstream and downstream discharges at point P, q_d^A and q_u^B are respectively downstream and upstream discharges at points A and B, A is the cross section of the pipe and H^P refers to the piezometric head at point P. Vaporous cavitation occurs when $H^P \le H_v - z\sin\gamma$, in which the gauge vapour head H_v is defined by:

$$H_v = \frac{p_v}{\rho_f g} \tag{1.17}$$

In addition, when a cavity collapses, its volume $\forall_v$ becomes zero or negative and the calculation returns to the standard water hammer equations. The change of the vapour cavity volume at a computational section is expressed in terms of the difference in discharges as:

$$\Delta\forall_v = \int_t^{t+2\Delta t} \left(q_d^P - q_u^P \right) dt \tag{1.18}$$

The numerical spikes are usually controlled with integration of Eq. (1.18). In a staggered grid, which is preferred to the normal rectangular grid [36], Eq. (1.18) becomes:

$$\forall_v^P = \forall_v^Q + 2\Delta t \left[\psi \left(q_d^P - q_u^P \right) + (1-\psi) \left(q_d^Q - q_u^Q \right) \right] \tag{1.19}$$

where ψ is a weighting factor normally equal to 0.5 for the trapezoidal rule. Numerical computations were performed by taking ψ in the range from 0.5 to 1 and concluded that if ψ is close to 1, pressure spikes are reduced [43 and 53]. Furthermore, the DVCM gives reasonably accurate results when the number of reaches is restricted so that the ratio of maximum cavity size to reach volume should be below 10 % [36].

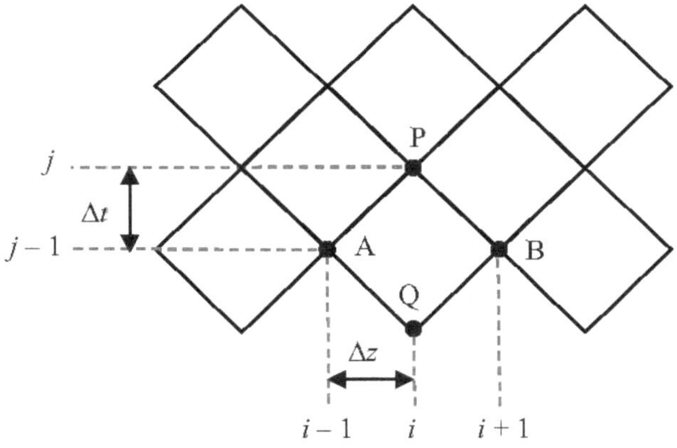

Fig. (1.13). Staggered grid (SG) of the MOC [52].

Although its simplicity, the DVCM has some weakness. First, internal boundary condition allows, subjectively, vapour cavities to form only at computational sections. The results are strongly influenced by the location of these sections. Secondly, to avoid the prediction of negative cavity volumes or negative absolute pressure, artificial restrictions are imposed, which produces large pressure peaks. Thirdly, since the cavity volume and the mass transfer are ignored at each computation section, then this model becomes insufficient to describe cavitation process. Finally, the discontinuity in the velocity of the liquid imposed at each computational section leads to two different values for the liquid velocity. The discrepancy increases with the cavitation severity and by decreasing the number of reaches. However, a high number of reaches implies large discontinuity, which leads to additional complication of the model.

One of the most realistic and effective methods used to approve the DVCM consists in allowing free gas in the cavity. Such method allowed the elaboration of the DGCM. Gas release is a physical phenomenon that occurs in hydraulic systems, like cooling water systems, long pipelines with high points, oil pipelines, sewage water lines, aviation fuel lines, *etc*. Once a gas, such as the air move through a free surface of a liquid, a certain amount of this gas is absorbed by the liquid. Entrapped gases are released when the pressure drops in the pipeline. If a cavity forms, it may be assumed that released gas stays in the cavity and does not immediately dissolve following a pressure rise. In contrast to vapour release, which takes only a few microseconds, the time for gas release is in the order of several seconds [6].

Valuable information about various models used to simulate column separation may be obtained in [24, 36 and 43]. The Discrete Vapour Cavity Model (DVCM), the Discrete Gaz Cavity Model (DGCM) and the Generalized Interface Vaporous Cavitation Model (GIVCM) are analysed and compared in [36].

The Classical DGCM

The DGCM has been used for modelling column separation over the last 30 years but not as widely as the DVCM although this latter is considered as limit-case of the DGCM. This model considers free gas volumes to simulate distributed free gas [6]. The entrained air effect on column separation was detailed by Estrada in 2007 [54], but earlier works had been developed in [6, 55, 56, 57 and 58]. Cavities are concentrated at the computational sections whereas pure liquid is assumed to remain in each computational reach [56 and 58]. A quantity of free gas is introduced at each computational section. Gas volumes at each computational section expanded and contracted with respect to the pressure according to an isothermal perfect gas law. This model exhibited dispersion of the wave front during rarefaction waves and steepening of the wave front for compressive waves. The DGCM was later validated against experimental results in [6, 36].

One of the most realistic and effective methods used to approve the DVCM consists in allowing free gas in the cavity. Such method allowed the elaboration of the

DGCM. Once a gas, such as the air move through a free surface of a liquid, a certain amount of this gas is absorbed by the liquid. Entrapped gases are released when the pressure drops in the pipeline. If a cavity forms, it may be assumed that released gas stays in the cavity and does not immediately dissolve following a pressure rise. In contrast to vapour release, which takes only a few microseconds, the time for gas release is in the order of several seconds [6].

The DGCM utilizes free gas volumes to simulate distributed free gas. In fact, the DVCM is assumed to be a specific case of the DGCM which is more general [22]. The DGCM allows cavities to be concentrated at computational sections, whereas pure liquid is assumed to remain in each computational reach [58]. In addition, a quantity of free gas is introduced at each computational section where gas volumes expanded and contracted as for isothermal perfect gas.

Considering a mixture volume $\forall_m$ of a pipeline, containing a gas volume $\forall_g$ and a liquid volume, the void fraction α defined by:

$$\alpha = \frac{\forall_g}{\forall_m} \tag{1.20}$$

is assumed to be constant along the pipeline. Isothermal behaviour of free gas can be a reasonable assumption in case of small void fractions [58]. The ideal gas law allows the expression of the void fraction with respect to the absolute partial pressure p_g^* and the absolute temperature T of the gas as:

$$\alpha = \frac{M_g R_g T}{p_g^* \forall_m} \tag{1.21}$$

where M_g is the mass of free gas and R_g is the specific gas constant. The DGCM considers a pipeline with concentrated gas volumes at computational sections. The amount of free gas concentrated at each section is determined by coalescing the distributed gas from the adjacent reach to a point [58]. The volume of fluid mixture in each adjacent reach is taken to be constant. By taking α_0 a void fraction at given reference pressure p_{g0}^*, Eq. (1.21) gives:

$$\alpha_0 = \frac{M_g R_g T}{p_{g0}^* \forall_m} \tag{1.22}$$

and subsequently the gas volume becomes:

$$\forall_g = \frac{\alpha_0 p_{g0}^* \forall_m}{p_g^*} \tag{1.23}$$

The Dalton's Law states that the total absolute pressure p^* at the computational section is the sum of the absolute gas pressure p_g^* and the vapour pressure p_v^*

$$p^* = p_g^* + p_v^* \tag{1.24}$$

which implies:

$$p_g^* = p^* - p_v^* = p - p_v \tag{1.25}$$

where p and p_v are respectively the total gauge pressure at the computational section and the vapour gauge pressure. By using the piezometric head instead of the pressure, it follows:

$$p_g^* = \rho_f g\left(H - Z - H_v\right) \tag{1.26}$$

Subsequently the expression of the gas volume $\forall_g$ is derived by using a staggered grid (SG) of characteristics (Fig. **4.1**). By considering a computational point P, $\forall_g$ is performed with respect of the piezometric head H at the point P

$$\forall_g^P = \frac{\alpha_0 p_{g0}^* \forall_m}{\rho g\left(H^P - Z - H_v\right)} \tag{1.27}$$

Wylie considered vaporous cavitation occurring for $\alpha_0 \leq 10^{-7}$ [58]. For this order of void fraction, there is only a small change in pressure wave-speed, even at low pressures.

The computational sections are treated as fixed internal boundary conditions. The computations of the DGCM as well as the DVCM is assumed to be simple, since one requires just a small modification on the compatibility equations derived from the water hammer reference model. The SG (Fig. **1.13**) is preferred for calculation using the MOC. The compatibility equations of the reference model along the characteristic directions C_f^+ and C_f^- are likely written as in [58].

$$C_f^+: \quad H^P - H^A + B\left(q_u^P - q_d^A\right) + Rq_u^P\left|q_d^A\right| = 0 \tag{1.28}$$

$$ZC_f^-: \quad H^P - H^B - B\left(q_d^P - q_u^B\right) - Rq_d^P\left|q_u^B\right| = 0 \tag{1.29}$$

with B is the pipeline impedance defined as:

$$B = \frac{C_f}{gA} \tag{1.30}$$

and R is its resistance coefficient given by:

$$R = \frac{f\Delta z}{2gDA^2} \tag{1.31}$$

The quantities $Rq_u^P\left|q_d^A\right|$ and $Rq_d^P\left|q_u^B\right|$ denote steady friction terms defined by the Darcy-Weisbach formula where f is the friction coefficient. Noting that if UF is considered, UF models can be used as described in [52, 59]. Wylie used a simple form to write Eqs. (4.27) and (4.28) namely [58].

$$C_f^+: \quad H^P = C_P - B_P Q_u^P \tag{1.32}$$

$$C_f^-: \quad H^P = C_M + B_M q_d^P \tag{1.33}$$

in which the constants C_P, B_P, C_M, B_M are:

$$C_P = H^A + Bq_d^A \quad \textbf{(1.34)}$$

$$B_P = B + R\left|q_d^A\right| \quad \textbf{(1.35)}$$

$$C_M = H^B - Bq_u^B \quad \textbf{(1.36)}$$

$$B_M = B + R\left|q_u^B\right| \quad \textbf{(1.37)}$$

Noting that if cavitation does not occur, the upstream and downstream discharges of each point of the computation grid are equal, and the pressure head is:

$$H^P = \frac{C_P B_M + C_M B_P}{B_P + B_M} \quad \textbf{(1.38)}$$

and the discharge is:

$$q^P = \frac{C_P - C_M}{B_P + B_M} \quad \textbf{(1.39)}$$

If gaseous cavitation occurs, H^P is an unknown that should be calculated, and one needs a fourth equation for calculation. The continuity equation can be used for such purpose as:

$$\frac{d\forall_g}{dt} = q_d^P - q_u^P \quad \textbf{(1.40)}$$

In a SG of characteristics (Fig. 1.13), the integration of Eq. (1.40) yields [58].

$$\forall_g^P = \forall_g^Q + 2\Delta t\left[\psi\left(q_d^P - q_u^P\right) + (1-\psi)\left(q_d^Q - q_u^Q\right)\right] \quad \textbf{(1.41)}$$

where ψ is the weighting factor already used for the classical DVCM. Eqs. (1.27), (1.28), (1.29) and (1.41) lead to the expression of the piezometric head at the point P [58].

$$H^P = \frac{1}{4}\left\{ -B_1 + 2(Z + H_v) + \left[\left(B_1 + 2(Z + H_v)\right)^2 + \frac{4Bp_0^*\alpha_0\forall_m}{\psi \Delta t \rho g} \right]^{1/2} \right\} \quad \textbf{(1.42)}$$

with

$$B_1 = -C_P - C_M + \left[\forall_m/(2\Delta t) + (1-\psi)\left(q_d^Q - q_u^Q\right) \right] B/\psi \quad \textbf{(1.43)}$$

Distributed Cavitation Models

Kalkwijk and Kranenburg defined the *bubble model* in 1971 and then in 1973 [60, 61]. Two approaches were proposed. The first one is based on the dynamic behaviour of vaporous bubbles whereas the second distinguishes between the water hammer zone and the cavitation zone where the wave-speed is reduced to zero. The bubble model allows the liquid pressure in the cavitation zone to be maintained equal to the vapour pressure p_v. When the cavitation zone stops growing, a choc wave forms at the interface separating the monophasic zone and the diphasic zone

and occupies the cavitating zone. This phenomenon was described with the continuity equation and the momentum equation.

In a thesis work at the university of Delft, in 1971, Kranenburg presented the effect of dissolved gas on the cavitating zones in pipelines [62]. Later, in 1974, he mentioned that the MOC cannot be used because the wave-speed is pressure dependent [63]. This dependence is mainly due to the dissolved gas. To simplify the model, the bubble flow is assumed to overly the whole pipe. The Lax-Wendroff finite difference method (FDM) was used by Kranenburg in 1972 to simulate the model [64]. In 1999, Hadj Taïeb used the Henry's law for gas release to construct the vaporous cavitation bubble [47]. His study was based on the diphasic flow assumption. The mixture was characterised by its mass density defined with respect of the vapour mass-fraction. Later, Haj Taïeb and Hadj Taïeb studied the bubble model by using the vapour void-fraction to express the mass density of the mixture in 2007 [65].

EXPERIMENTAL INVESTIGATION

In this subsection, some water hammer experiments carried out on copper pipes (elastic) and polyethylene pipe (viscoelastic) are presented where water is used as fluid. The mass density and the Young's modulus of elasticity of copper are constant for all tests and respectively equal to $8900\ \text{kg.m}^{-1}$ and 10^5 GPa. Also, the Poisson coefficient of copper pipe is of about 0.3. for all tests, transient was caused by an instantaneous closing downstream valve.

Experiment of Haj Taïeb

Haj Taïeb carried out experimental tests on water hammer and cavitation at the FMI (Fluid Mechanic Institute) of Toulouse in 1973. A standard reservoir-pipe-valve system was used. The copper pipe is of a length L = 35.7 m, a diameter D = 19.6 mm, a thickness e = 1 mm and a friction coefficient f = 0.03. The first case corresponds to the low speed $V_0 = 0.095\ \text{m.s}^{-1}$. The pressure history diagram (Fig. **1.14**) allows the calculation of the period T and the pressure wave-speed $C_f = 1150\ \text{m.s}^{-1}$. The second task for the Haj Taïeb experiment is the high-speed case $V_0 = 0.56\ \text{m.s}^{-1}$ for which cavitation occurs (Fig **1.14**). The collapse of the first cavity leads to a pressure rise higher than the Joukowsky pressure rise.

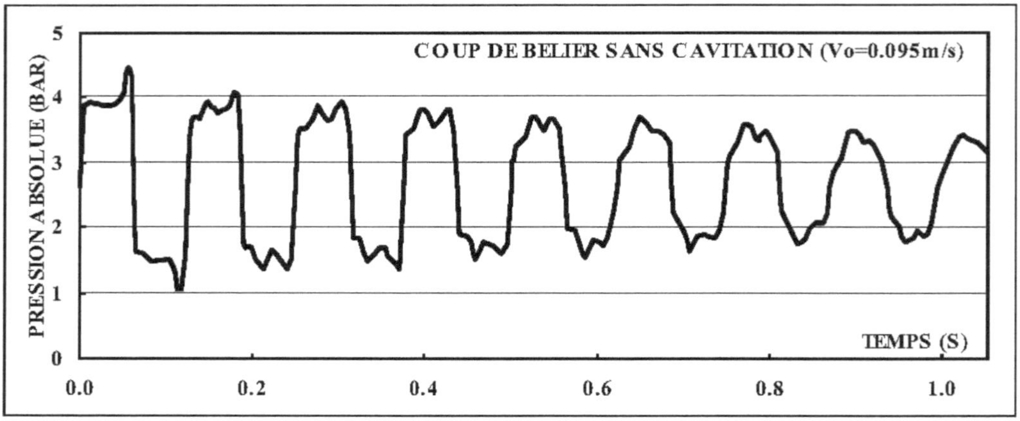

(Fig. 1.14) contd.....

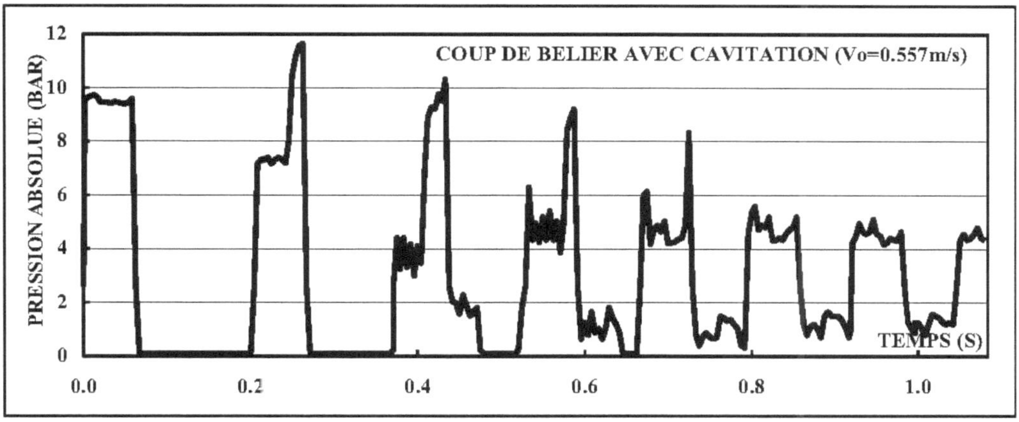

Fig. (1.14). Pressure history at the valve in the experiment of Haj Taïeb. Top: low-speed case, bottom: high-speed case [20].

Experiment of Simpson

The experiment of Simpson (Fig. **1.15**) was carried out at The University of Michigan. Major components of the experimental apparatus include an upstream reservoir, a 36 m long and 19.05 mm inside diameter copper pipe, a downstream one-quarter turn ball valve, and a downstream reservoir. The pipe-wall thickness is of 1.588 mm. The pipeline has a right-angle bend located 12.5 meters from the upstream end. Fig. (**1.16**) shows a HGL variation test for the cavitation case. Further spectral analysis of pressure was performed to estimate the wave-speed.

Experiment of Bergant and Simpson

The experimental test shown in Fig. (**1.17**) was performed by Bergant and Simpson in 1995 [66]. The apparatus is composed of a copper straight 37.23 m long sloping pipeline of 22.1 mm internal diameter and 1.6 mm wall thickness connecting two pressurized tanks and fulfilled by a demineralized water. The pipe slope has a constant value of 3.27 °. The properties of water at 20 °C are the following: density $\rho = 998$ Kg.m^{-3}, bulk modulus of elasticity K = 2.19 GPa, kinematic viscosity $\nu' = 1.01 \times 10^{-6}$ m^2.s^{-1} and absolute vapour pressure $p_v = 2.34$ kPa . Copper used for the construction of the pipe is characterized by a Young's modulus $E = 120 \pm 5$ GPa and a Poisson's ratio $\nu = 0.34$. The calculation of the anchor coefficient leads to $c_1 = 1.02$ (Bergant and Simpson, 1995).

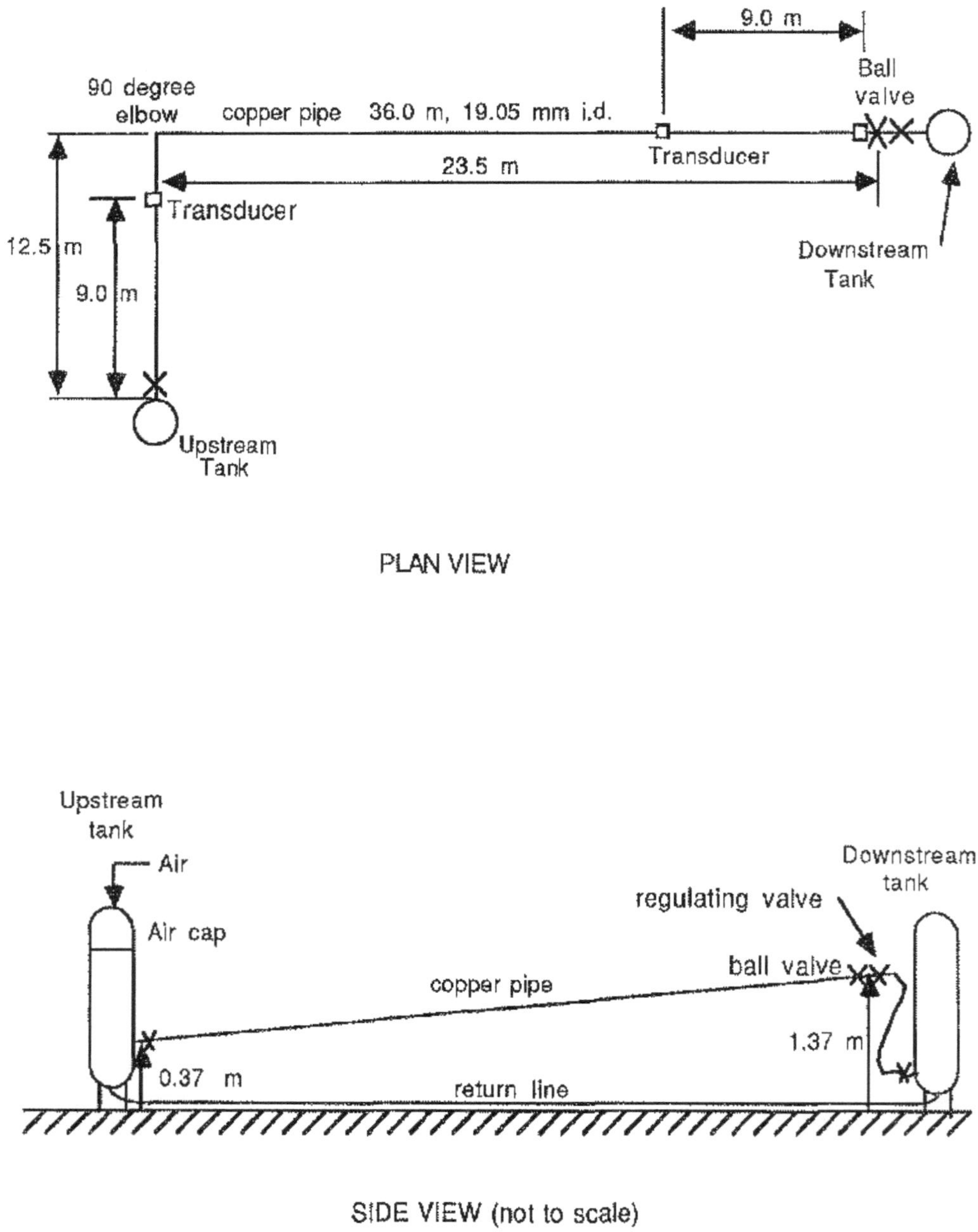

Fig. (1.15). Layout of the experiment of Simpson [24].

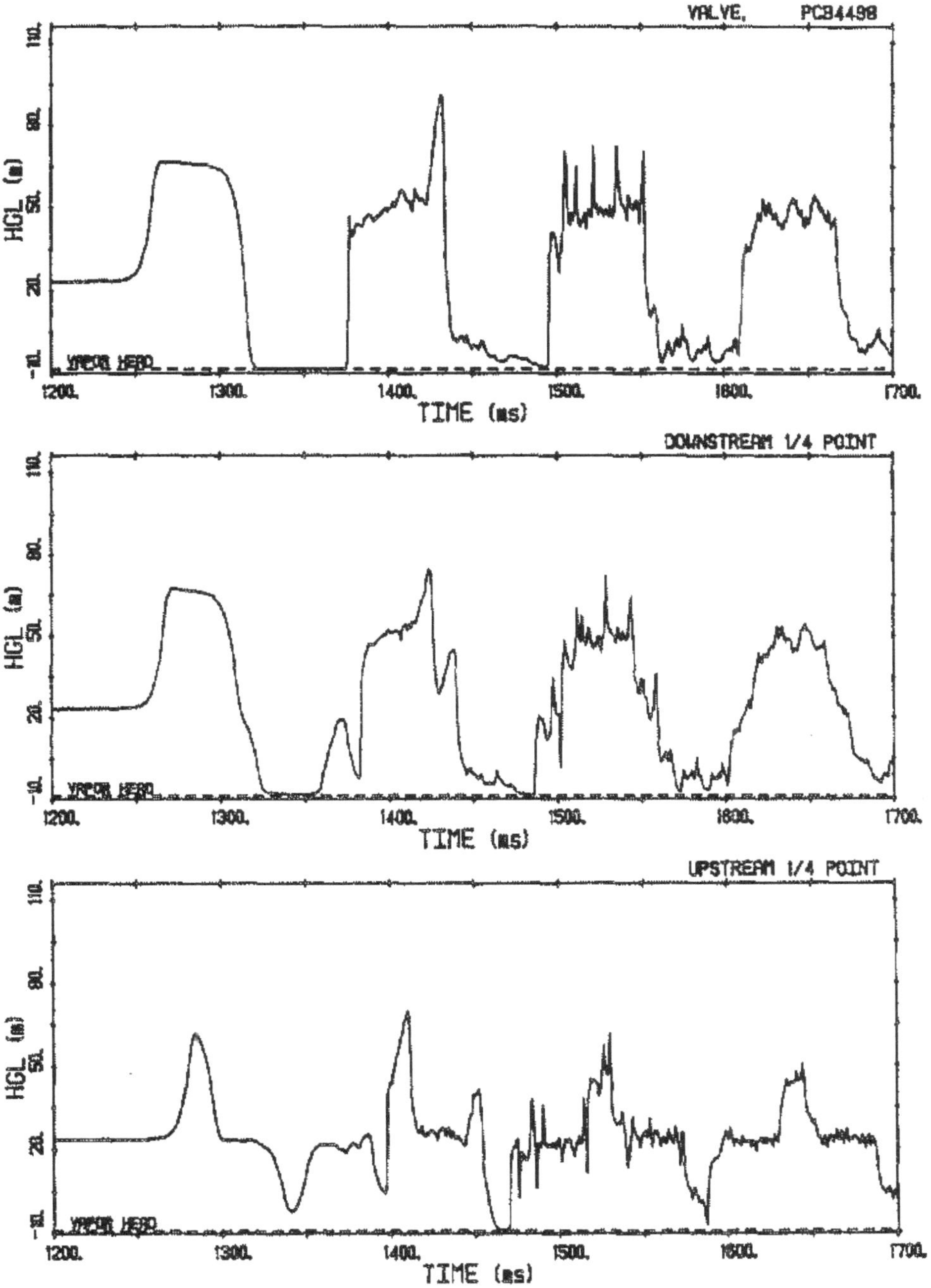

Fig. (1.16). Hydraulic Grade Line variation with time for cavitation case in the experiment of Simpson, $V_0 = 0.332\ \mathrm{m.s}^{-1}$ and $H_{res} = 23.41$ m [24].

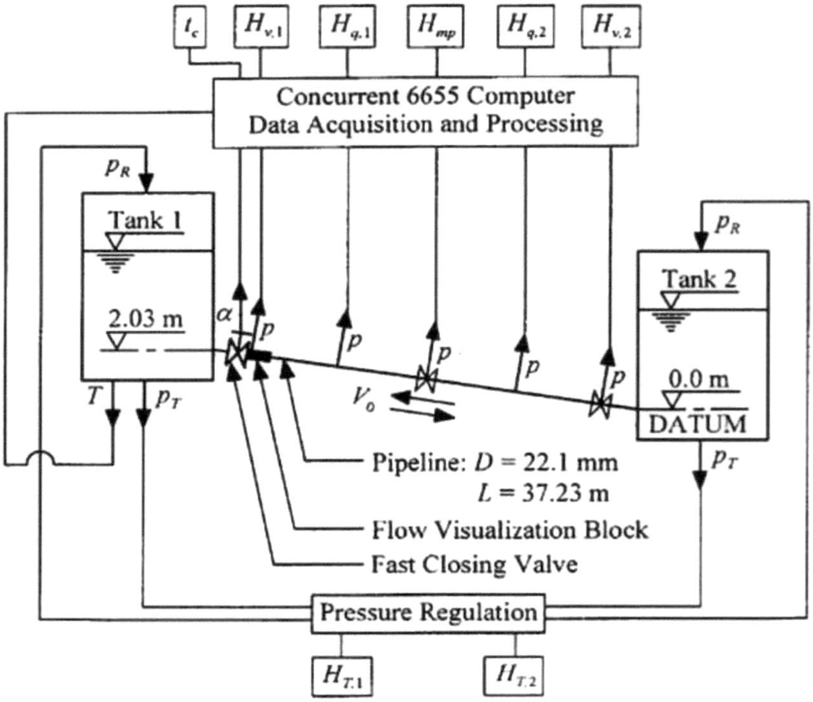

Fig. (1.17). Experimental apparatus layout of the experiment of Bergant and Simpson [36].

A pressure control system for maintaining a specified pressure in each of the tanks enables the simulation of transients in either an upward or a downward sloping pipeline. The pressure in both tanks may be regulated from 20 to 620 kPa to simulate low-head or medium-head hydraulic systems. However, the net water volume in both tanks and the capacity of the air compressor limit the maximum steady-state velocity to 1.5 m.s^{-1} and maximum operating pressure (pressure head) in each tank to 400 kPa (40 m).

Water hammer events including column separation in the experimental apparatus are initiated by rapid closure of a ball valve. The valve can be located at either end of the pipeline adjacent to either tank or at the midpoint of the pipeline; thus, simulation of various types of pipe configurations can be performed. For a downstream end closing valve, column separation can occur at the valve. Depending on the slope of the pipe, a region of distributed vaporous cavitation may or may not occur along an extended length of the pipeline. For an upstream end closing valve, the hydraulic grade line (HGL) on the downstream side of the valve may drop to the vapour pressure head resulting in a localised cavity forming adjacent to the valve. Again, distributed vaporous cavitation may or may not occur along an extended length of the pipeline. Pressure transducers are located at five equidistant points along the pipeline, as close as possible to the endpoints.

After an initial steady state is established, a transient event is initiated by a rapid valve closure. The water hammer method pipe flow velocity is computed from the Joukowsky pressure head rise or pressure head drop resulting from a fast closure of the valve. For instance, Fig. (**1.18**) shows HGL at the valve at the downstream end in case of upstream sloping pipe with an initial velocity of $0.40\ \mathrm{m.s^{-1}}$.

The accuracy of the measurement was compared for two pressure transducer technologies: the Druck pressure transducer (Fig. **1.18a**) and the Kistler pressure transducer (Fig. **1.18b**). The pressure records show that the latter exhibits less pressure oscillations, but the timing is the same for the two technologies.

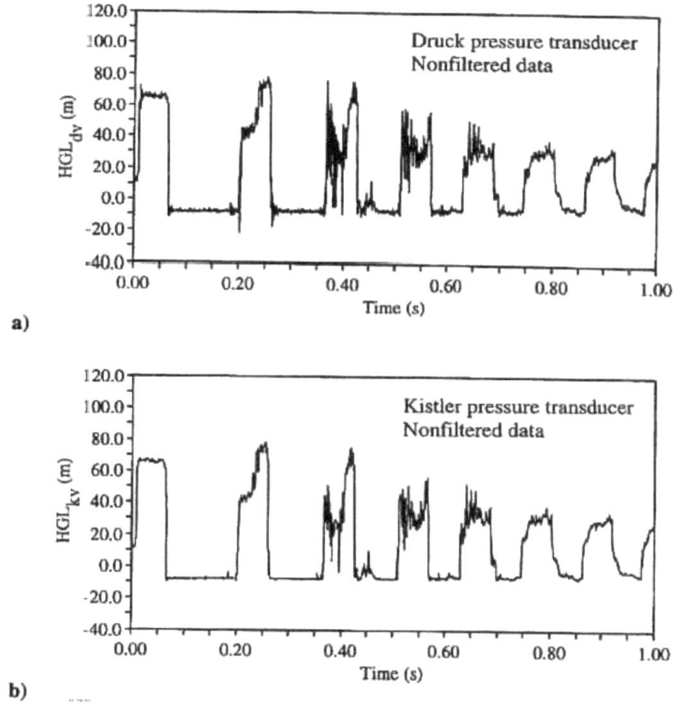

Fig. (1.18). Comparison of nonfiltered data of HGLs at the fast-closing valve in the experiment of Bergant and Simpson (upward sloping pipe, valve at the downstream end, $V_0 = 0.40\ \mathrm{m.s^{-1}}$)[66].

Experiment of Stoinov and Covas

An experimental facility with a single pipeline system was assembled at the Imperial College of London [1 et 67]. This facility (Fig. **1.19**) was designed for the analysis and testing of novel leak detection techniques based on the generation of transient events in the fluid system. It was used to collect pressure and strain data for the analysis of the mechanical behaviour of polyethylene pipe.

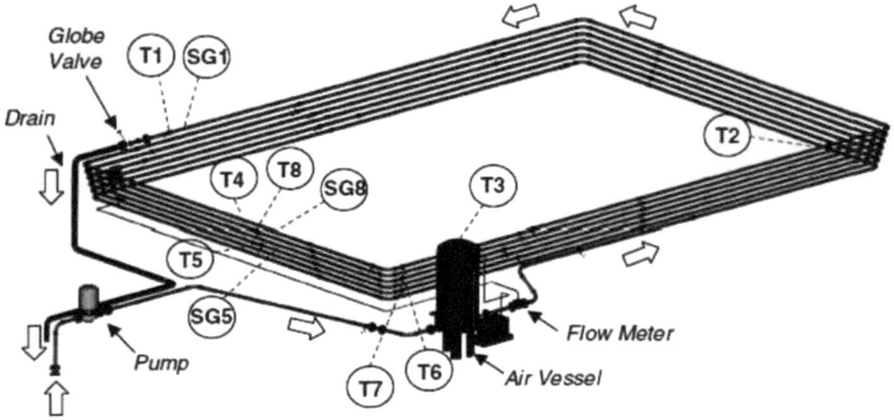

Fig. (1.19). Imperial College experimental facility of the experiment of Stoinov and Covas. [67].

The pipeline is made of high-density polyethylene SDR11 PE100 NP16, with 63 mm nominal diameter (ND) and 6.2 mm wall thickness. The total length of the pipeline is 277 m (length between the vessel and the downstream valve). Pipe sections are electro-fused and rigidly fixed to a vertical wall with plastic brackets, 1m spaced along its length, and with metal frames at the elbows, to restrain the pipe from any axial movement. The pipe-rig includes a centrifugal pump and a pressurized tank with 750 l, at the upstream end, and a globe valve to control the flow and to generate transient events, at the downstream end. The globe valve discharges directly to a free surface flow drainage pipe. Various data sets were collected.

Fig. (**1.19**) shows the location and distance from the upstream end of the measurement sites. Strain gauges were installed in the circumferential direction of the pipe to measure the radial displacement of the pipe cross-section. The axial strain was not measured because the pipe was restrained from moving axially and, in these circumstances, axial strain is negligible compared to the circumferential strain.

Fig. (**1.20**) shows pressure and circumferential strain at transducers T1, T3, T5 and T8 localized at 271 m, 0 m, 116.42 m, and 196.93 m, respectively.

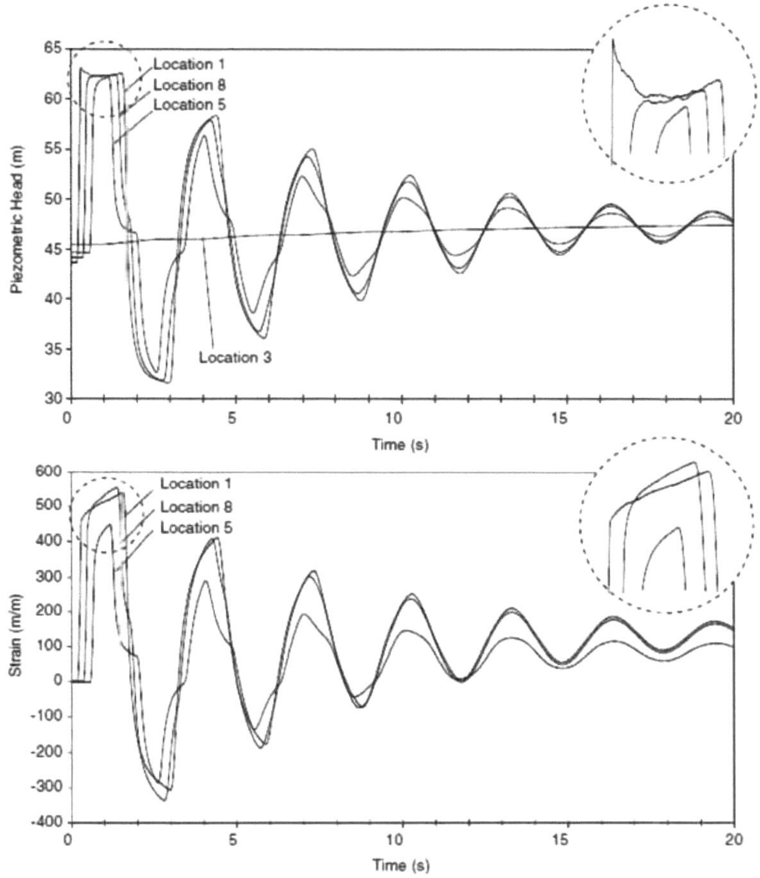

Fig. (1.20). Transient data collected at pressure transducers T3, T5, T8 and T1, and strain gauges SG5, SG8 and SG1 in the experiment of Stoinov and Covas ($Q_0 = 1.0$ l.s^{-1} and $T = 20$ °C) [67].

Experiment of Carriço

The experiment of Carriço (Fig. **1.21**) was assembled at Instituto Superior Técnico of Lisbon, in 2008. This experiment was carried out on a high-density polyethylene (HDPE) pipes [68]. The experimental facility was composed of a single transmission pipeline with a total length of 203 m and inner diameter of 44 mm. This pipeline was connected to an air vessel at the upstream end and to a free discharge outlet into a constant water level at the downstream end. A ball valve used to interrupt the flow was installed immediately downstream the air vessel. The air vessel was used to keep the upstream pressure constant as an elevated reservoir.

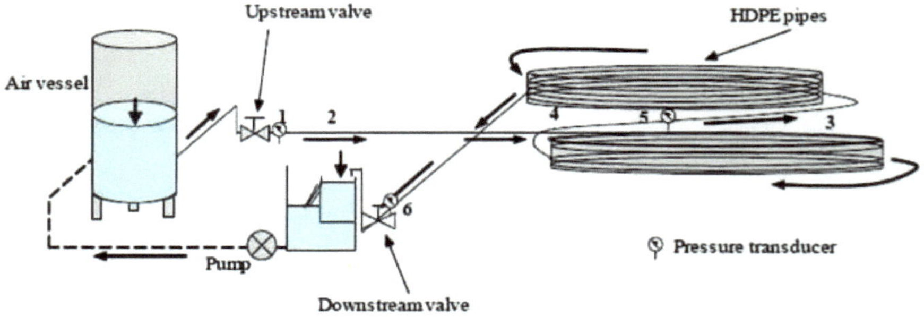

Fig. (1.21). Experimental set-up of Carriço [68].

Transient pressure data have been collected using pressure transducers located at four pipe sections with a frequency of 500 Hz: at the air vessel, downstream the ball valve (section 1), at the middle of the pipeline (section 5) and at downstream end of the pipeline (section 6).

The experiment of Carriço presented two cases of fluid transient: the non-cavitating case and the cavitating case. The fast closure of the upstream end ball valve caused transient for the two cases. The non-cavitating case was characterized by a steady state flow rate $Q_0 = 2.72 \text{ l.s}^{-1}$ *i.e.*, Re = 80000 and a pressure wave speed estimated at $C_f = 315 \text{ m.s}^{-1}$. However, in the cavitating case, the flow rate is greater $Q_0 = 4.0 \text{ l.s}^{-1}$, *i.e.*, Re = 120000 whereas the pressure wave speed is lesser and estimated at $C_f = 250 \text{ m.s}^{-1}$.

Column separation in the HDPE was simulated by use of different models taking into account pipe-wall viscoelasticity [68]. Fig. (**1.22**) illustrates the efficiency of the DGCM in prediction of column separation at section 1.

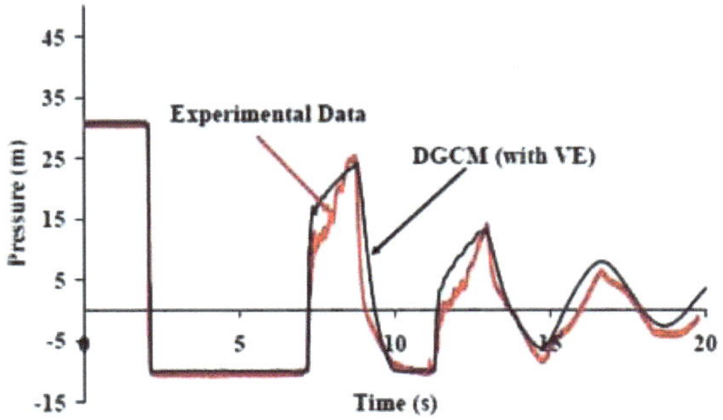

Fig. (1.22). Prediction of pressure at section 1 in the experiment of Carriço [68].

CONCLUSION

In this chapter, an extensive overview of the several research areas was carried out. Issues like hydraulic transients, water hammer, cavitation, fluid-structure interaction (FSI), mathematical modelling of fluid transients and further experimental investigation was reviewed. The physical phenomena were described with emphasis on vaporous and gaseous cavitation and column separation in pipelines. The mathematical models of water hammer and cavitation were introduced without details. A development of these mathematical models will be the subject of the next chapter. Numerous experimental tests from the literature were presented. They will be used in chapters 5 and 6 to validate the proposed models.

CHAPTER 2

Mathematical Development

Abstract: This chapter gives a detailed description of the mathematical development proposed for the study. The elasticity theory and the general theory of beams are strongly present whereas some assumptions are adopted to simplify the analysis.

Keywords: Creep; Elastic pipelines; Fluid-structure interaction; Poisson coupling; Viscoelasticity.

INTRODUCTION

The mathematical development of the water hammer models is based on two equations : (i) the continuity equation and (ii) the momentum equation. Both of them represent an independent mechanical law. When applied to the fluid in the pipe, *e.g.*, water, the two equations lead to establishing the partial differential equation (PDE) of the fluid, which constitutes the classical water hammer model described in Chapter 1. However, if fluid structure interaction (FSI) is involved, then two other equations are added leading to the four-equation model. The procedure is applied to water hammer and cavitation in both elastic and viscoelastic pipelines.

WATER HAMMER IN ELASTIC PIPELINES

Introduction

Water hammer in elastic pipelines (metallic ad concrete pipelines) is considered in this section. The pipe of inner diameter D, thickness e and inclination γ contains a moving fluid. At each section of the pipe, one can define the steady state velocity V and the pressure p. The four-equation model (4EM) is used to determine these two variables for the fluid and the axial velocity and the axial stress for the pipe. The formulation of the 4EM is based on two basic principles: the continuity principle and the energy conservation principle. The FSI in liquid-filled piping systems is modelled by extended water hammer theory for the fluid and the Timoshenko beam theory and elasticity theory (Hooke's law) for the pipe [2, 45 and

Abdelaziz Ghodhbani, Ezzeddine Haj Taïeb, Mohsen Akrout & Sami Elaoud

69]. Thus, two governing equations for the fluid are coupled to two governing equations for the pipe by means of boundary conditions, representing the contact between the fluid and the pipe-wall at the interface.

Assumptions

Schwartz introduced the following assumptions in order to simplify the analysis [2 and 16]:

- Transient flow is one-dimensional: pressure p and velocity V are uniform on each section A of the pipeline.
- Convective terms are neglected since the velocity V is very smaller than the pressure wave celerity C_f. This assumption is called *acoustic approximation.*
- The pressure waves propagate at low frequencies in axial direction; radial inertia effect is neglected.
- The fluid is homogeneous, isotropic and has linearly elastic behavior.
- The material of the pipe is isotropic and homogeneous, and it has a quasi-rigid linear elastic behavior (metals and concrete).
- The pipe has a circular thin-walled section.
- Cavitation does not occur: the fluid pressure remains above the vapour pressure.

Fluid Dynamics

The cylindrical coordinate system is considered. For an infinitesimal fluid volume ϑ, the continuity equation can be derived in its local form [70]:

$$\frac{\partial \rho_f}{\partial t} + V_r \frac{\partial \rho_f}{\partial r} + \frac{V_\varphi}{r}\frac{\partial \rho_f}{\partial \varphi} + V_z \frac{\partial \rho_f}{\partial z} + \rho_f div\vec{V} = 0 \qquad (2.1)$$

The equation of state is:

$$\frac{\partial \rho_f}{\partial t} = \frac{\rho_f}{K}\frac{\partial p}{\partial t} \qquad (2.2)$$

in which K is the bulk modulus of compressibility of the fluid. Since the fluid is homogeneous, its density does not depend on space variables, *i.e.*, $\partial \rho_f / \partial r = \partial \rho_f / \partial z = \partial \rho_f / \partial \varphi = 0$. After neglecting the circumferential part of the velocity [2], Eqs. (2.1) and (2.2) lead to the continuity equation of the fluid.

$$\frac{1}{K}\frac{\partial p}{\partial t}+\frac{\partial V_z}{\partial z}+\frac{1}{r}\frac{\partial}{\partial r}\left(rV_r\right)=0 \tag{2.3}$$

For the same infinitesimal volume ϑ, the equation of motion can also be written in a local form as:

$$\rho_f \frac{d\vec{V}}{dt}=-\overrightarrow{grad}\,p+\overrightarrow{div\bar{\bar{\tau}}}+\vec{F} \tag{2.4}$$

with $\vec{F}$ is the outside force per unit volume and $\bar{\bar{\tau}}$ denotes the viscosity stress tensor given by:

$$\bar{\bar{\tau}}=\left(\mu'-\frac{2}{3}\mu\right)div\vec{V}\bar{\bar{I}}+2\mu\bar{\bar{D}} \tag{2.5}$$

with μ and μ' are respectively dynamic and volumetric viscosity coefficients, $\bar{\bar{I}}$ is the identity matrix and $\bar{\bar{D}}$ is the strain tensor defined as:

$$\bar{\bar{D}}=\frac{1}{2}\left(\overline{\overline{grad\vec{V}}}+\left(\overline{\overline{grad\vec{V}}}\right)^t\right) \tag{2.6}$$

Eqs. (2.5) and (2.6) can be introduced in Eq. (2.4) and lead to the Navier-Stokes equation.

$$\rho_f \frac{d\vec{V}}{dt}=\rho_f\vec{F}-\overrightarrow{grad}\,p+\mu\Delta\vec{V}+\left(\mu'+\frac{1}{3}\mu\right)\overrightarrow{grad}\left(div\vec{V}\right) \tag{2.7}$$

in which the coefficient μ' is equal to zero. The left-hand side of Eq. (2.7) is a particular derivative that can be developed as:

$$\rho_f \frac{d\vec{V}}{dt}=\rho_f\left(\frac{\partial\vec{V}}{\partial t}+\left(\overline{\overline{grad\vec{V}}}\right)\vec{V}\right) \tag{2.8}$$

in which:

$$\frac{\partial \vec{V}}{\partial t} = \left(\frac{\partial V_r}{\partial t} - \frac{V_\varphi^2}{r} \right) \overrightarrow{u_r} + \left(\frac{\partial V_\varphi}{\partial t} - \frac{V_r V_\varphi}{r} \right) \overrightarrow{u_\varphi} + \frac{\partial V_z}{\partial t} \overrightarrow{u_z} \quad \textbf{(2.9)}$$

Since the circumferential component is neglected, Eq. (2.9) becomes:

$$\frac{\partial \vec{V}}{\partial t} = \frac{\partial V_r}{\partial t} \overrightarrow{u_r} + \frac{\partial V_z}{\partial t} \overrightarrow{u_z} \quad \textbf{(2.10)}$$

The vector $\vec{F} = \rho_f \vec{g}$ in Eq. 2.7 is vertical, *i.e.* $\vec{F} = F_r \overrightarrow{u_r} + F_z \overrightarrow{u_z}$. Since the radial inertia is ignored, the above vector becomes:

$$\vec{F} = \rho_f g \sin \gamma \overrightarrow{u_z} \quad \textbf{(2.11)}$$

In addition, the pressure p is assumed to be independent to the angle φ , which implies:

$$\overrightarrow{grad p} = \frac{\partial p}{\partial r} \overrightarrow{u_r} + \frac{\partial p}{\partial z} \overrightarrow{u_z} \quad \textbf{(2.12)}$$

Furthermore, the Laplacian operator of the vector $\vec{V}$ is:

$$\Delta \vec{V} = \overrightarrow{grad}\left(div \vec{V} \right) - \overrightarrow{rot}\left(\overrightarrow{rot \vec{V}} \right) \quad \textbf{(2.13)}$$

Since the flow is assumed to be isochoric, *i.e*, $div \vec{V} = 0$ and subsequently, Eq. (2.13) is simplified because $\overrightarrow{grad}\left(div \vec{V} \right) = 0$. Moreover, by ignoring the circumferential component V_φ and the angle φ , the development of Eq. (2.13) in the cylindrical coordinate system $\left(\overrightarrow{u_r}, \overrightarrow{u_\varphi}, \overrightarrow{u_z} \right)$ leads to [70]:

$$\begin{pmatrix} \frac{1}{r}\frac{\partial}{\partial r}\left(r \frac{\partial V_r}{\partial r} \right) + \frac{\partial^2 V_r}{\partial z^2} - \frac{V_r}{r^2} \\ 0 \\ \frac{1}{r}\frac{\partial}{\partial r}\left(r \frac{\partial V_z}{\partial r} \right) + \frac{\partial^2 V_z}{\partial z^2} \end{pmatrix} = \begin{pmatrix} \frac{\partial}{\partial z}\left(\frac{\partial V_r}{\partial z} - \frac{\partial V_z}{\partial r} \right) \\ 0 \\ -\frac{1}{r}\frac{\partial}{\partial r}\left[r \left(\frac{\partial V_r}{\partial z} - \frac{\partial V_z}{\partial r} \right) \right] \end{pmatrix} \quad \textbf{(2.14)}$$

By assuming $\partial V_r / \partial z = 0$, the axial component of $\Delta \vec{V}$ is derived:

$$\Delta V_z = \frac{1}{r}\frac{\partial}{\partial r}\left(r\frac{\partial V_z}{\partial r} \right) \tag{2.15}$$

Since the convective terms are neglected, *i.e.*, $\partial V_z / \partial z = 0$ and $V_r\, \partial V_r / \partial r = 0$, Eq. (2.8) can be simplified and the equations of motion respectively in radial and axial direction are obtained.

$$\rho_f \frac{\partial V_r}{\partial t} + \frac{\partial p}{\partial r} = 0 \tag{2.16}$$

$$\rho_f \frac{\partial V_z}{\partial t} + \frac{\partial p}{\partial z} = \rho_f g \sin\gamma + \frac{\mu}{r}\frac{\partial}{\partial r}\left(r\frac{\partial V_z}{\partial r} \right) \tag{2.17}$$

The present study will be limited to the axial direction. The continuity (2.3) and the equation of motion (2.17) present nonlinear terms that can lead to integration problems. To avoid this problem, each equation is multiplied by r and then integrated on r between 0 and R. The continuity equation and momentum equation are, respectively,

$$\frac{1}{K}\frac{\partial p}{\partial t} + \frac{\partial V_z}{\partial z} + \frac{2}{R} V_r \big|_{r=R} = 0 \tag{2.18}$$

$$\rho_f \frac{\partial V_z}{\partial t} + \frac{\partial p}{\partial z} = \rho_f g \sin\gamma + \frac{2\mu}{R}\frac{\partial V_z}{\partial r}\Big|_{r=R} \tag{2.19}$$

It is worth noting that the integration is obtained by considering the assumption of one-D flow.

Pipe Dynamics

In this study, the pipe motion is assumed to be isochoric. The continuity equation is:

$$\frac{d\rho_p}{dt} + \rho_p div\vec{u} = 0 \tag{2.20}$$

with ρ_p and $\vec{u}$ are respectively the mass density and the axial velocity of the pipe-wall. The second component of Eq. (2.20) can be neglected since the deformation of the pipe is elastic. This leads to:

$$\frac{d\rho_p}{dt} = \frac{\partial \rho_p}{\partial t} + \vec{u}\,\overrightarrow{grad}\rho_p = 0 \tag{2.21}$$

By assuming that the pipe material is homogenous and constant in time, it can be derived:

$$\overrightarrow{grad}\rho_p = 0 \text{ and } \frac{\partial \rho_p}{\partial t} = 0 \tag{2.22}$$

The above condition leads to the ignorance of the continuity equation in the dynamic analysis of the pipe.

The local Euler's fundamental equations of motion in the cylindrical coordinate system $\left(\overrightarrow{u_r}, \overrightarrow{u_\varphi}, \overrightarrow{u_z}\right)$ with origin O placed on the central axis of the pipe are respectively the momentum and the moment of momentum equations [70].

$$\rho_p \frac{d\vec{u}}{dt} = \rho_p\left(\frac{\partial \vec{u}}{\partial t} + \vec{u}.\overline{\overline{grad}}\vec{u}\right) = div\overline{\overline{\sigma}} + \vec{f} \tag{2.23}$$

$$\rho_p \frac{d}{dt}\left(\overrightarrow{OM} \times \vec{u}\right) = \overrightarrow{OM} \times \overrightarrow{div\overline{\overline{\sigma}}} + \overrightarrow{OM} \times \vec{f} + \overline{\overline{\overline{\eta}}} : \overline{\overline{\sigma}} \tag{2.24}$$

where $\overline{\overline{\sigma}}$ denotes the Cauchy's stress tensor (symmetric tensor), $\vec{f}$ is the body-force density for the pipe, *i.e.* $\vec{f} = \rho_p \vec{g}$, the vector $\overrightarrow{OM} \times \vec{u}$ is the kinetic moment of the point M relatively to the origin O and $\overline{\overline{\overline{\eta}}}$ is the orientation tensor. In long wave approximation, the kinetic momentum is ignored. In addition, all static moment (bending moment and torsion moment) are neglected. As a result, the moment of momentum equation (2.24) is ignored. In addition, the circumferential component of the momentum equation (2.23) is neglected. Therefore, two momentum equations are retained (respectively radial and axial directions).

$$\rho_p \frac{\partial \dot{u}_r}{\partial t} + \rho_p \dot{u}_r \frac{\partial \dot{u}_r}{\partial r} + \rho_p \dot{u}_r \frac{\partial \dot{u}_z}{\partial z} = \frac{\partial \sigma_r}{\partial r} + \frac{\partial \tau_{rz}}{\partial z} + \frac{\sigma_r}{r} - \frac{\sigma_\varphi}{r} + f_r \quad \textbf{(2.25)}$$

$$\rho_p \frac{\partial \dot{u}_z}{\partial t} + \rho_p \dot{u}_r \frac{\partial \dot{u}_r}{\partial r} + \rho_p \dot{u}_r \frac{\partial \dot{u}_z}{\partial z} = \frac{\partial \sigma_z}{\partial z} + \frac{\partial \tau_{rz}}{\partial r} + \frac{\tau_{rz}}{r} + f_z \quad \textbf{(2.26)}$$

As mentioned in 1.3, the convective terms should be neglected since the axial velocity of the pipe is very smaller than the stress wave-speed C_p. In addition, the radial body-force density is also omitted. Moreover, the shear stress τ_{rz} is assumed to be independent of the axial coordinate *z*. Thus, eqs (2.25) and (2.26) becomes, respectively:

$$\rho_p \frac{\partial \dot{u}_r}{\partial t} = \frac{1}{r} \frac{\partial (r\sigma_r)}{\partial r} - \frac{\sigma_\varphi}{r} \quad \textbf{(2.27)}$$

$$\rho_p \frac{\partial \dot{u}_z}{\partial t} = \frac{\partial \sigma_z}{\partial z} + \frac{1}{r} \frac{\partial (r\tau_{rz})}{\partial r} + \rho_p g \sin\gamma \quad \textbf{(2.28)}$$

To more develop the above equations, Eq. (2.27) is multiplied by *r*, integrated with respect to *r* from *R* to *R* + *e* and then divided by *e*. Eq. (2.28) is multiplied by $2\pi r$, integrated with respect to *r* from *R* to *R* + *e* and then divided by $\pi(2R+e)e$. The results are respectively in radial and axial direction as follows [2]:

$$\overline{\sigma_\varphi} = \frac{R+e}{e} \sigma_r \big|_{r=R+e} - \frac{R}{e} \sigma_r \big|_{r=R} \quad \textbf{(2.29)}$$

$$\rho_p \frac{\partial \overline{\dot{u}_z}}{\partial t} = \frac{\partial \overline{\sigma_z}}{\partial z} - \frac{2R}{(2R+e)e} \tau_{rz} \big|_{r=R} + \rho_p g \sin\gamma \quad \textbf{(2.30)}$$

with $\sigma_r \big|_{r=R+e}$ is equal to outside pressure usually equal to the barometric pressure. The variables $\overline{\dot{u}_z}$, $\overline{\sigma_\varphi}$ and $\overline{\sigma_z}$ are respectively averaged values of $\dot{u}_z$, σ_φ and σ_z obtained as:

$$\overline{\dot{u}_z} = \frac{1}{\pi(2R+e)e} \int_R^{R+e} 2\pi r \dot{u}_z dr \quad \textbf{(2.31)}$$

$$\overline{\sigma_\varphi} = \frac{1}{e}\int_R^{R+e} \sigma_\varphi dr \quad (2.32)$$

$$\overline{\sigma_z} = \frac{1}{\pi(2R+e)e}\int_R^{R+e} 2\pi r\sigma_z dr \quad (2.33)$$

and the averages values $\overline{\sigma_r}$ and $\overline{\overline{\sigma_\varphi}}$ are:

$$\overline{\sigma_r} = \frac{1}{\pi(2R+e)e}\int_R^{R+e} 2\pi r\sigma_r dr \quad (2.34)$$

$$\overline{\overline{\sigma_\varphi}} = \frac{1}{\pi(2R+e)e}\int_R^{R+e} 2\pi r\sigma_\varphi dr \quad (2.35)$$

The average value $\overline{\sigma_\varphi}$ has its expression different from the other averages values because σ_φ depends on the radius *r*. Noting that Eqs. (2.29) and (2.30) present three unknowns: $\dot{u}_z$, σ_φ and σ_z. The third equation can be obtained thanks to the elasticity laws [2].

Fluid-Pipe Coupling

The fluid and the pipe are coupled at the interface $r = R$. The governing equations of both fluid and pipe are considered. The dynamic equilibrium at the interface is described by the following conditions [2]:

$$\sigma_r|_{r=R} = -p|_{r=R} \quad (2.36)$$

$$\dot{u}_r|_{r=R} = V_r|_{r=R} \quad (2.37)$$

$$\tau_{rz}|_{r=R} = \mu\frac{\partial V_z}{\partial r}|_{r=R} = -\tau_0 \quad (2.38)$$

The shear stress τ_{rz} can be obtained after writing the static equilibrium using the Darcy-Weisbach formula. A portion *L* of a pipeline of inner radius *R* into which a fluid of density ρ_f is flowing at a velocity V_z is considered. The axial velocity of the pipe-wall is $\dot{u}_z$. For steady state, the pressure change Δp through the portion L can be written with respect to the relative velocity $V_{rel} = V_z - \dot{u}_z$ as:

$$\Delta p = \frac{f\rho_f L}{4R}(V_z - \dot{u}_z)|V_z - \dot{u}_z| = \frac{f\rho_f L}{4R}V_{rel}|V_{rel}| \qquad \textbf{(2.39)}$$

with f is the friction coefficient and $|V_z - \dot{u}_z|$ is used to distinguish the sign of Δp. the quasi-static equilibrium of the pipe-wall in steady state condition is:

$$\pi R^2 \Delta p = 2\pi R L \tau_0 \qquad \textbf{(2.40)}$$

Then, Eqs. (2.38) to (2.40) lead to:

$$\tau_{rz}|_{r=R} = -\frac{f\rho_f}{8}V_{rel}|V_{rel}| \qquad \textbf{(2.41)}$$

The Four-Equation Model

For the fluid, after introducing the boundary condition (2.37) into (2.18), the continuity equation becomes

$$\frac{1}{K}\frac{\partial p}{\partial t} + \frac{\partial V_z}{\partial z} + \frac{2}{R}\dot{u}_r|_{r=R} = 0 \qquad \textbf{(2.42)}$$

The time derivative of Eq. (A.9) can be incorporated into Eq. (2.42), which leads to:

$$\frac{1}{K}\frac{\partial p}{\partial t} + \frac{\partial V_z}{\partial z} + \frac{2}{E}\frac{\partial}{\partial t}\left(\sigma_\varphi|_{r=R} - \nu\sigma_z|_{r=R} - \nu\sigma_r|_{r=R}\right) = 0 \qquad \textbf{(2.43)}$$

After that, the circumferential stress $\sigma_\varphi|_{r=R}$ can be defined thanks to Eq. (B.18) as:

$$\sigma_\varphi|_{r=R} = \left(\frac{R}{e} + \frac{e}{2R+e}\right)p \qquad \textbf{(2.44)}$$

and the term $\sigma_z|_{r=R}$ can be replaced by $\overline{\sigma_z}$ [2]. Finally, by taking account of the boundary condition (2.36), Eq. (2.43) leads to the final formulation of the continuity equation of the fluid.

$$\frac{\partial V_z}{\partial z}+\left[\frac{1}{K}+\frac{2}{E}\left(\frac{R}{e}+\frac{e}{2R+e}+\nu\right)\right]\frac{\partial p}{\partial t}-\frac{2\nu}{E}\frac{\partial \overline{\sigma_z}}{\partial t}=0 \quad \textbf{(2.45)}$$

Eqs. (2.38) and (2.41) lead to the development of the momentum equation of the fluid in axial direction.

$$\frac{\partial V_z}{\partial t}+\frac{1}{\rho_f}\frac{\partial p}{\partial z}=-\frac{f}{4R}V_{rel}\left|V_{rel}\right|+g\sin\gamma \quad \textbf{(2.46)}$$

For the pipe, Eqs. (A.18) and (A.19) give:

$$\frac{1}{\nu}\sigma_z-\frac{E}{\nu}\frac{\partial u_z}{\partial z}=\sigma_\varphi+\sigma_r \quad \textbf{(2.47)}$$

which implies:

$$\frac{1}{\nu}\overline{\sigma_z}-\frac{E}{\nu}\frac{\partial \overline{u_z}}{\partial z}=\overline{\overline{\sigma_\varphi}}+\overline{\sigma_r} \quad \textbf{(2.48)}$$

By using Eqs. (B.17), (B.18), (2.34) and (2.35), it follows:

$$\overline{\overline{\sigma_\varphi}}+\overline{\sigma_r}=\frac{2R^2}{(2R+e)e}p \quad \textbf{(2.49)}$$

Hence, Eq. (2.48) becomes:

$$\frac{1}{\nu}\overline{\sigma_z}-\frac{E}{\nu}\frac{\partial \overline{u_z}}{\partial z}-\frac{2R^2}{(2R+e)e}p=0 \quad \textbf{(2.50)}$$

The time derivative of Eq. (2.50) gives:

$$\frac{\partial \overline{\dot{u}_z}}{\partial z}-\frac{1}{E}\frac{\partial \overline{\sigma_z}}{\partial t}+\frac{2\nu R^2}{E(2R+e)e}\frac{\partial p}{\partial t}=0 \quad \textbf{(2.51)}$$

By introducing Eq. (2.40) into the momentum equation (2.30), it follows:

$$\frac{\partial \overline{\dot{u}_z}}{\partial t} - \frac{1}{\rho_p}\frac{\partial \overline{\sigma_z}}{\partial z} = \frac{f \rho_f R V_{rel}\left|V_{rel}\right|}{4\rho_p(2R+e)e} + g\sin\gamma \quad \textbf{(2.52)}$$

The four equations (2.45), (2.46), (2.51) and (2.52) form the four-equation model. However, it is commonly to modify Eq. (2.45) using Eq. (2.51) and the result is:

$$\frac{\partial V_z}{\partial z} + \left[\frac{1}{K} + \frac{2}{E}\left(\frac{R}{e} + \frac{e}{2R+e} + \nu - \frac{R}{e}\nu^2\right)\right]\frac{\partial p}{\partial t} - 2\nu\frac{\partial \overline{\dot{u}_z}}{\partial z} = 0 \quad \textbf{(2.53)}$$

Based on the thin-wall pipe assumption *i.e.*, $e/R \ll 1$, the averaged variables can be replaced by the local variables. Subsequently, the following approximation can be used:

(i) in Eq. (2.45): $R/e + e/(2R+e) = R/e$;

(ii) in Eq. (2.51): $R/\left[e(1+e/D)\right] = R/e$;

(iii) in Eq. (2.52): $R/\left[e/(2R+e)\right] = 1/(2e)$

These approximations and Eq. (1.10) allow the following modification in Eq. (2.53).

$$\left[\frac{1}{K} + \frac{2}{E}\left(\frac{R}{e} + \frac{e}{2R+e} + \nu - \frac{R}{e}\nu^2\right)\right]\frac{\partial p}{\partial t} \approx \left[\frac{1}{K} + \frac{2R}{Ee}\left(1-\nu^2\right)\right]\frac{\partial p}{\partial t} = \frac{1}{\rho_f C_f^2}\frac{\partial p}{\partial t} \quad \textbf{(2.54)}$$

After replacing the averaged variables by the local variables, the four-equation model appears:

$$\frac{\partial V_z}{\partial t} + \frac{1}{\rho_f}\frac{\partial p}{\partial z} = -\frac{f}{4R}V_{rel}\left|V_{rel}\right| + g\sin\gamma \quad \textbf{(2.55)}$$

$$\frac{\partial V_z}{\partial z} + \frac{1}{\rho_f C_F{}^2}\frac{\partial p}{\partial t} - 2\nu\frac{\partial \dot{u}_z}{\partial z} = 0 \quad \textbf{(2.56)}$$

$$\frac{\partial \dot{u}_z}{\partial t} - \frac{1}{\rho_p}\frac{\partial \sigma_z}{\partial z} = \frac{f}{8e}\frac{\rho_f}{\rho_p} V_{rel}\left|V_{rel}\right| + g\sin\gamma \tag{2.57}$$

$$\frac{\nu R}{Ee}\frac{\partial p}{\partial t} + \frac{\partial \dot{u}_z}{\partial z} - \frac{1}{E}\frac{\partial \sigma_z}{\partial t} = 0 \tag{2.58}$$

Another way to write the four-equation model consists in replacing the gauge pressure p by the piezometric head H according to the following equation:

$$H = \frac{p}{\rho_f g} + Z \tag{2.59}$$

In case of the reservoir-pipe-valve system, by assuming the axis z oriented from the reservoir up to the valve (Figs. **2.1** and **2.2**) and the inclination γ of the pipe positive regardless the pipe sloping, the head Z can be expressed as:

$$Z = \begin{cases} -z\sin\gamma + h_0, & \text{for downward sloping pipes} \\ z\sin\gamma + h_0, & \text{for upward sloping pipes} \end{cases} \tag{2.60}$$

with h_0 is an arbitrary datum.

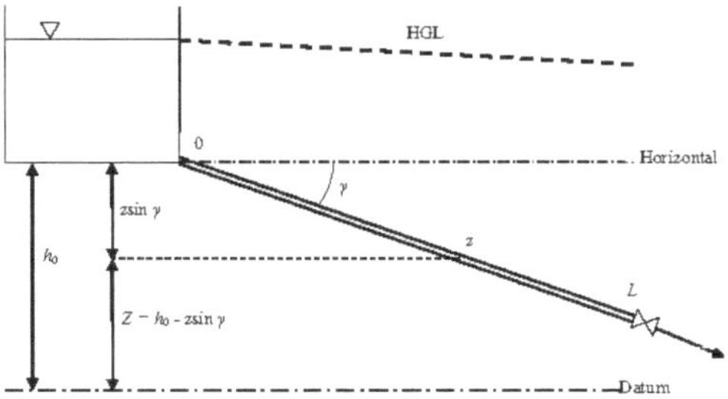

Fig.(2.1). Piezometric head layout for a downward sloping pipe.

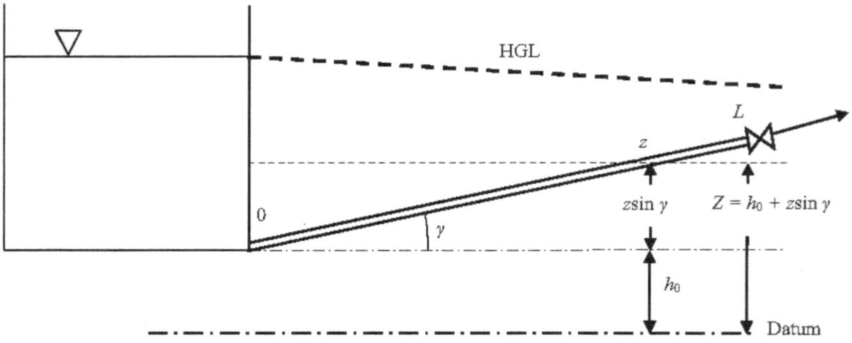

Fig. (2.2). Piezometric head layout for an upward sloping pipe.

For instance, $h_0 = L\sin\gamma$ in case of downward sloping pipes if the datum is taken at the valve, and $h_0 = 0$ for upward sloping pipes if the datum is taken at the reservoir delivery. Noting that the present study considers the case of downward sloping pipes because the axial body-force f_z is positive. The final governing equations of the four-equation model will be the same regardless the pipe sloping.

The use of the expression (2.59) leads to:

$$\frac{\partial V_z}{\partial t} + g\frac{\partial H}{\partial z} = -\frac{f}{4R}V_{rel}\left|V_{rel}\right| \tag{2.61}$$

$$\frac{\partial V_z}{\partial z} + \frac{g}{C_f^2}\frac{\partial H}{\partial t} - 2\nu\frac{\partial \dot{u}_z}{\partial z} = 0 \tag{2.62}$$

$$\frac{\partial \dot{u}_z}{\partial t} - \frac{1}{\rho_p}\frac{\partial \sigma_z}{\partial z} = \frac{f}{8e}\frac{\rho_f}{\rho_p}V_{rel}\left|V_{rel}\right| + g\sin\gamma \tag{2.63}$$

$$\rho_f g\frac{\nu R}{Ee}\frac{\partial H}{\partial t} + \frac{\partial \dot{u}_z}{\partial z} - \frac{1}{E}\frac{\partial \sigma_z}{\partial t} = 0 \tag{2.64}$$

The Four-Equation Friction Model

The four-equation friction model (4EFM) was proposed by Ghodhbani and Haj Taïeb [59]. The model is obtained by incorporating a given friction model (such as the Zielke model, the Vardy-Brown model, the Brunone model, *etc.*) into the 4EM. This allows this latter to take account of unsteady friction (UF) which is an important damping parameter for water hammer. The steady state friction term given by the Darcy-Weisbach formula is replaced by the head loss h_f which is obtained by the superposition of steady and unsteady parts. The relative velocity V_{rel} is used instead the fluid velocity V_z.

$$h_f = \frac{f}{2gD} V_{rel} \left| V_{rel} \right| + \frac{16\nu}{gD^2} \left(\frac{\partial V_{rel}}{\partial t} * W_0 \right)(t) \tag{2.65}$$

in which the weighting function W_0 depends on the flow condition (laminar or turbulent). The incorporation of the head loss h_f given by the formula (2.65) into the governing equations leads to (the inner diameter D is used instead of the inner radius R).

$$\frac{\partial V_z}{\partial t} + g \frac{\partial H}{\partial z} = -g h_f \tag{2.66}$$

$$\frac{\partial V_z}{\partial z} + \frac{g}{C_f^2} \frac{\partial H}{\partial t} - 2\nu \frac{\partial \dot{u}_z}{\partial z} = 0 \tag{2.67}$$

$$\frac{\partial \dot{u}_z}{\partial t} - \frac{1}{\rho_p} \frac{\partial \sigma_z}{\partial z} = g \frac{\rho_f}{\rho_p} \frac{D}{4e} h_f + g \sin \gamma \tag{2.68}$$

$$\rho_f g \frac{\nu D}{2Ee} \frac{\partial H}{\partial t} + \frac{\partial \dot{u}_z}{\partial z} - \frac{1}{E} \frac{\partial \sigma_z}{\partial t} = 0 \tag{2.69}$$

WATER HAMMER IN VISCOELASTIC PIPES

Introduction

Plastic pipes (such as polyethylene) are being increasingly used in hydraulic plants because of their high mechanical and chemical properties. Viscoelasticity can mainly attenuate the pressure fluctuations by increasing the dispersion of the travelling wave. The viscoelastic approach allows strain to be decomposed into instantaneous elastic strain and retarded viscoelastic strain. The former is included in the wave speed, whereas the latter is included as an additional term in the continuity equation [25]. The responses of typical viscoelastic material are shown in Fig. (**2.3**). To describe the viscoelastic linear response, the Kelvin-Voight model can be used. This model consists in associating in parallel a springer and a damper (Fig. **2.4**). According to this model, the stress σ is related to the strain ε by the following [71] represented by the damper. The generalized Kelvin-Voight model shown by Fig. (**2.4c**) is the most reliable representation.

$$\sigma = E_0\varepsilon + \mu\dot{\varepsilon} \tag{2.70}$$

with E_0 is the Young's modulus of elasticity represented by the spring, μ is the viscosity.

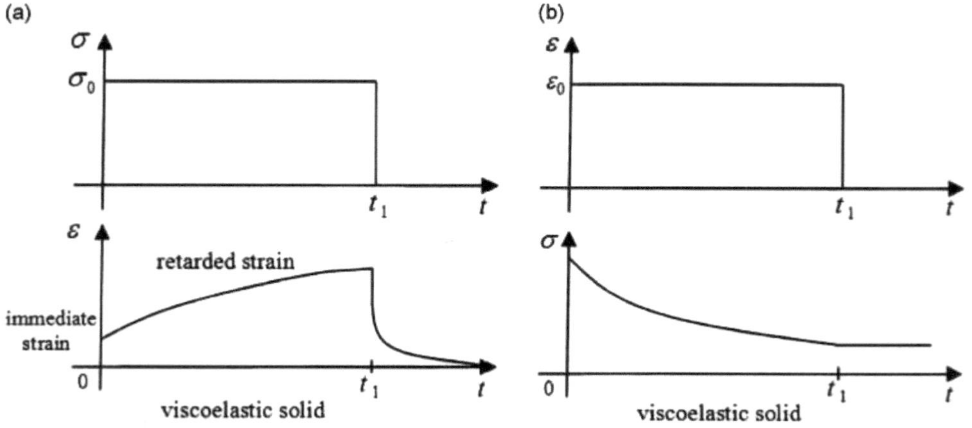

Fig. (2.3). Responses of typical viscoelastic material (**a**) creep test and (**b**) relaxation test [72].

When subjected to an instantaneous stress σ, the linear viscoelastic material does not respond according to Hook's Law, but it has an instantaneous elastic strain ε_e and a retarded viscous strain ε_r [73].

$$(\varepsilon)_t = \varepsilon_e + (\varepsilon_r)_t \tag{2.71}$$

The Classical Viscoelastic Model

The Boltzmann's superposition principle established that for small strains, the combination of stresses acting independently results in strains that can be added linearly. Hence, the total strain ε generated by a continuous application of stress σ is:

$$\varepsilon = \sigma J_0 + \sigma * \frac{\partial J}{\partial t} \tag{2.72}$$

where J_0 is the instantaneous creep-compliance given by $J_0 = 1/E_0$ for linearly elastic materials, J is the creep-compliance function of time t, and "*" denotes convolution. Eq. (2.72) can be developed as [73]:

$$(\varepsilon)_t = (\sigma)_t J_0 + \int_0^t (\sigma)_{t-s} \frac{\partial (J)_s}{\partial s} ds \tag{2.73}$$

in which $(J)_s$ is the creep-compliance function at time s. The Stieltjes convolution notation "*d" is in general defined as:

$$(F * dG)_t = (F)_t G_0 + \int_0^t (F)_{t-s} \frac{\partial (G)_s}{\partial s} ds \tag{2.74}$$

Thus, Eq. (2.74) can be written as:

$$(\varepsilon)_t = (\sigma * dJ)_t \tag{2.75}$$

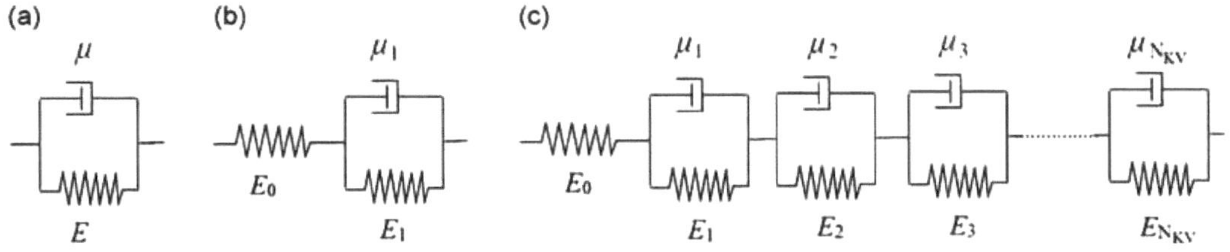

Fig. (2.4). Mechanical representation of a viscoelastic solid: (**a**) one Kelvin Voigt element, (**b**) three-parameter Kelvin Voigt model and (**c**) generalised Kelvin Voigt model [71].

For a pipe of inner diameter D and thickness e, it is assumed that the pipe material is (i) homogeneous an isotropic, (ii) its behaviour is linear viscoelastic for small strains, (iii) Poisson's ratio is constant so that the material behaviour depends only on the creep-functions, the circumferential stress can be approximated as:

$$\sigma_\varphi = \frac{D\Delta p}{2e} = \frac{D}{2e}\left[(p)_t - p_0\right] \tag{2.76}$$

with $(p)_t$ is the pressure at time t and p_0 is the pressure at time t = 0. The total circumferential strain is:

$$(\varepsilon_\varphi)_t = \frac{D_0}{2e_0}\left[(p)_t - p_0\right]J_0 + \int_0^t \frac{(D)_{t-s}}{2(e)_{t-s}}\left[(p)_{t-s} - p_0\right]\frac{\partial (J)_s}{\partial s}\,ds \qquad \textbf{(2.77)}$$

where the subscript 0 denotes the variable at time $t = 0$. The first part of Eq. (2.77) corresponds to the elastic strain and the integral part to the retarded strain. For the generalized Kelvin-Voight model, the creep-compliance function $(J)_t$ can be expressed as:

$$(J)_t = J_0 + \sum_{k=1}^{N} J_k\left(1 - e^{-t/\tau_k}\right) \qquad \textbf{(2.78)}$$

with J_k is the creep-compliance of the spring of the Kelvin-Voight k-element defined with respect to modulus of elasticity E_k by $J_k = 1/E_k$, τ_k is the retardation time of the dashpot of the k-element given by $\tau_k = \mu_k / E_k$ and μ_k is the viscosity of the dashpot.

By neglecting the convective terms, it can be assumed that:

$$V_z \ll C_f \text{ and } \dot{u}_z \ll C_p \qquad \textbf{(2.79)}$$

and by using the relation between the cross-sectional area A and the total circumferential strain ε_φ given by $dA/dt = 2Ad\varepsilon_\varphi / dt$, the Reynolds transport theorem allows further development of the continuity equation (1.12) in which the velocity V is replaced by the discharge $Q = AV$ [73].

$$\frac{\partial H}{\partial t} + \frac{C_f^2}{gA}\frac{\partial Q}{\partial z} + 2\frac{C_f^2}{g}\frac{\partial \varepsilon_{\varphi.r}}{\partial t} = 0 \qquad \textbf{(2.80)}$$

The retarded strain is represented in the third term of Eq. (2.80), while the elastic strain is included in the pressure head derivative and in the elastic wave speed C_f. The compatibility equations corresponding to the continuity equation (2.80) and the momentum equation (1.13) which stay unchanged with the linear viscoelastic behaviour are:

$$\frac{dH}{dt} \pm \frac{C_f}{gA}\frac{dQ}{dt} + 2\frac{C_f^2}{g}\left(\frac{\partial \varepsilon_{\varphi.r}}{\partial t}\right) \pm C_f h_f = 0 \tag{2.81}$$

in which h_f is the friction losses head. The above equations are valid for straight characteristic directions.

The Four-Equation Viscoelastic Model

The water hammer viscoelastic model which is the application of the water hammer reference model to viscoelastic pipes is already introduced in the first chapter. For viscoelastic material, Eq. (1.17) is an alternative to represent the stress strain relation [71]. The formulation of the four-equation viscoelastic model (4EVEM) assumes constant Poisson's ratio ν. The axial strain in the cylindrical coordinate system is introduced using the Stieltjes convolution notation as [71].

$$\varepsilon_z = \sigma_z * dJ - \nu\left(\sigma_\varphi * dJ - \sigma_r * dJ\right) \tag{2.82}$$

Since the radial stress is neglected compared to the hoop stress, then Eq. (2.82) is reduced to:

$$\varepsilon_z = \sigma_z * dJ - \nu\sigma_\varphi * dJ \tag{2.83}$$

and similarly, the circumferential strain is:

$$\varepsilon_\varphi = \sigma_\varphi * dJ - \nu\sigma_z * dJ \tag{2.84}$$

Hence, Eqs. (2.83) and (2.84) are mainly used in deriving the governing equations for the 4EVEM. According to the viscoelastic behaviour of polymers, it can be stated that the application of sudden stress σ_0 to a viscoelastic beam results in strain that can be written as follows [71].

$$\varepsilon = \sigma_0 * dJ = \sigma_0 J_0 + \sigma_0 \sum_{k=1}^{N} J_k\left(1 - e^{-t/\tau_k}\right) \tag{2.85}$$

and subsequently one can write:

$$\varepsilon_\infty = \lim_{t\to\infty} \varepsilon = \sigma_0 \sum_{k=0}^{N} J_k \tag{2.86}$$

The 4EVEM simulates water hammer in viscoelastic pipes with FSI. As assumed for the previous study performed for the elastic 4EM, only axial vibration of the pipe is considered. The mechanical damping resulting from the retarded strain affects both fluid and structural responses. Viscoelastic damping is generally very larger than friction damping. Frictionless equations are considered in coupled modelling of fluid transients in viscoelastic pipelines [71]. In this study, UF is ignored, but SF is still considered. The first step is to consider the continuity equation (2.42) derived for the elastic model for which the radial displacement can be replaced according to the relation: $u_r = r\varepsilon_\varphi$. Consequently, Eq. (2.42) implies:

$$\frac{\partial V_z}{\partial z} + \frac{\rho_f g}{K}\frac{\partial H}{\partial t} + 2\frac{\partial \varepsilon_\varphi}{\partial t} = 0 \tag{2.87}$$

Then, the incorporation of Eq. (2.84) in Eq. (2.87) leads to:

$$\frac{\partial V_z}{\partial z} + \frac{\rho_f g}{K}\frac{\partial H}{\partial t} + 2\frac{\partial\left(\sigma_\varphi * dJ\right)}{\partial t} - 2\nu\frac{\partial\left(\sigma_z * dJ\right)}{\partial t} = 0 \tag{2.88}$$

Eq. (2.76) allows the use of the piezometric head change $\Delta H = H - H_0$ instead of the circumferential stress as:

$$\frac{\partial\left(\sigma_\varphi * dJ\right)}{\partial t} = \frac{D\rho_f g}{2e}\frac{\partial\left(\Delta H * dJ\right)}{\partial t} \tag{2.89}$$

Whereas the axial stress can be obtained thanks to Eq. (2.83).

$$\frac{\partial\left(\sigma_z * dJ\right)}{\partial t} = \frac{\partial \varepsilon_z}{\partial t} + \nu\frac{\partial\left(\sigma_\varphi * dJ\right)}{\partial t} \tag{2.90}$$

According to Eq. (2.76) and the definition of ε_z in the cylindrical coordinate system, it follows:

$$\frac{\partial(\sigma_z * dJ)}{\partial t} = \frac{\partial \dot{u}_z}{\partial z} + \nu \rho_f g \frac{D}{2e} \frac{\partial(\Delta H * dJ)}{\partial t} \tag{2.91}$$

Thus, the incorporation of Eqs. (2.89) and (2.91) into Eq. (2.88) leads to the continuity equation of the fluid in the viscoelastic pipe.

$$\frac{\partial V_z}{\partial z} + \frac{\rho_f g}{K} \frac{\partial H}{\partial t} - 2\nu \frac{\partial \dot{u}_z}{\partial z} + \left(1 - \nu^2\right) \rho_f g \frac{D}{e} \frac{\partial(\Delta H * dJ)}{\partial t} = 0 \tag{2.92}$$

The development of the Stieltjes convolution form of the term $\Delta H^* dJ$ leads to:

$$\Delta H * dJ = \left((H)_t - H_0\right) J_0 + \int_0^t \left((H)_{t-s} - H_0\right) \frac{\partial (J)_s}{\partial s} ds \tag{2.93}$$

and subsequently,

$$\frac{\partial(\Delta H * dJ)}{\partial t} = \frac{\partial H}{\partial t} J_0 + \frac{\partial}{\partial t} \left[\int_0^t \left((H)_{t-s} - H_0\right) \frac{\partial (J)_s}{\partial s} ds \right] \tag{2.94}$$

The second set of equations governing axial vibration of the pipe are derived from Eq. (2.91) in which:

$$\frac{\partial(\sigma_z * dJ)}{\partial t} = \frac{\partial \sigma_z}{\partial t} J_0 + \frac{\partial}{\partial t} \left[\int_0^t (\sigma_z)_{t-s} \frac{\partial (J)_s}{\partial s} ds \right] \tag{2.95}$$

As assumed for the hoop stress in Eq. (2.76), the axial stress can be also approximated as:

$$\sigma_z = \frac{D^2}{4e(D+e)} \Delta p = \frac{D^2}{4e(D+e)} \left[(p)_t - p_0\right] \approx \frac{D}{4e} \left[(p)_t - p_0\right] \tag{2.96}$$

Then, the incorporation of the expressions (2.94), (2.95) and (2.96) in Eq. (2.91) gives the first momentum equation of the vibrating pipe in axial direction.

Finally, the constitutive equations of the 4EVEM read as follows:

$$\frac{\partial V_z}{\partial t}+g\frac{\partial H}{\partial z}=-gh_f \quad (2.97)$$

$$\frac{\partial V_z}{\partial z}+\frac{g}{C_f^2}\frac{\partial H}{\partial t}-2\nu\frac{\partial \dot{u}_z}{\partial z}=\left(\nu^2-1\right)\frac{D}{e}\rho_f g\frac{\partial}{\partial t}\left(\int_0^t\left((H)_{t-s}-H_0\right)\frac{\partial (J)_s}{\partial s}ds\right) \quad (2.98)$$

$$\frac{\partial \dot{u}_z}{\partial t}-\frac{1}{\rho_p}\frac{\partial \sigma_z}{\partial z}=g\frac{\rho_f}{\rho_p}\frac{D}{4e}h_f+g\sin\gamma \quad (2.99)$$

$$\frac{\partial \dot{u}_z}{\partial z}-\frac{1}{E_0}\frac{\partial \sigma_z}{\partial t}+\frac{\nu\rho_f g}{E_0}\frac{D}{2e}\frac{\partial H}{\partial t}= \left(\nu-\frac{1}{2}\right)\frac{D}{2e}\rho_f g\frac{\partial}{\partial t}\left(\int_0^t\left((H)_{t-s}-H_0\right)\frac{\partial (J)_s}{\partial s}ds\right) \quad (2.100)$$

where the pressure wave-speed is defined as $C_f=\left\{\rho_f\left[1/K+\left(1-\nu^2\right)D/\left(eE_0\right)\right]\right\}^{-1/2}$ and the stress wave-speed is $C_p=\sqrt{E_0/\rho_p}$

INITIAL CONDITIONS

It can be stated that for types of the above mathematical models, the hydraulic system is in equilibrium until transient takes place. The initial velocity of the pipe is equal to zero. Initial conditions for all the previous models are obtained from the steady state conditions. By eliminating the time derivative in Eqs. (2.55) to (2.58), it follows:

$$\frac{dp}{dz}=-\frac{f\rho_f V_z\left|V_z\right|}{4R}+\rho_f g\sin\gamma \quad (2.101)$$

$$\frac{dV_z}{dz}=0 \quad (2.102)$$

$$\frac{d\dot{u}_z}{dz} = 0 \tag{2.103}$$

$$\frac{d\sigma_z}{dz} = -\frac{\rho_f f V_z |V_z|}{8e} - \rho_p g \sin\gamma \tag{2.104}$$

The integration of the above system with respect to the axial coordinate z gives:

$$p_0(z) = p_{res} - \left(\frac{f \rho_f V_{z0}^2}{4R} - \rho_f g \sin\gamma \right) z \tag{2.105}$$

$$V_{z0}(z) = V_{z0}(0) = V_{z0} \tag{2.106}$$

$$\sigma_{z0}(z) = \sigma_{z0}(0) - \left(\rho_p g \sin\gamma + f \rho_f \frac{V_{z0}^2}{8e} \right) z \tag{2.107}$$

$$\dot{u}_{z0}(z) = \dot{u}_{z0}(0) = 0 \tag{2.108}$$

with p_{res} is the pressure of the reservoir. Upstream the pipe, the initial axial stress can be calculated with respect to the initial pressure. Since the displacement at the reservoir is equal to zero, Eq. (A10) gives:

$$\sigma_z(0) = \nu\left(\sigma_\varphi(0) + \sigma_r(0)\right) \tag{2.109}$$

By taking account of Eq. (2.49), the expression (2.109) becomes:

$$\sigma_{z0}(0) = \frac{2\nu R^2}{(2R+e)e} p_{res} \approx \frac{\nu R}{e} p_{res} \tag{2.110}$$

The displacement at a given section z can be obtained by integrating Eq. (A10).

$$u_z = u_{z0}(0) + \frac{1}{E}\int\left(\sigma_z - \nu\left(\sigma_\varphi + \sigma_r\right)\right) dz \tag{2.111}$$

By assuming the initial axial displacement equal to zero at the reservoir, Eq. (2.111) leads to:

$$u_{z0}(z) = \frac{1}{E}\int\left[\sigma_{z0}(z) - \nu\left(\sigma_{\varphi 0}(z) + \sigma_{r0}(z)\right)\right]dz \qquad \textbf{(2.112)}$$

At the steady state, Eq. (2.49) reads:

$$\sigma_{\varphi 0}(z) + \sigma_{r0}(z) = \frac{2R^2}{(2R+e)e}p_0(z) \approx \frac{R}{e}p_0(z) \qquad \textbf{(2.113)}$$

Consequently, Eqs. (107) and (109) allow:

$$\sigma_{z0}(z) = \frac{\nu R}{e}p_{rés} - \left(\rho_p g \sin\gamma + \frac{\rho_f f V_{z0}^2}{8e}\right)z \qquad \textbf{(2.114)}$$

Finally, the initial displacement can be written:

$$u_{z0}(z) = \frac{z^2}{2E}\left[\left(\nu\rho_f - \rho_p\right)g\sin\gamma - \left(\frac{1}{2e} + \frac{\nu}{R}\right)\frac{f\rho_f V_{z0}^2}{4}\right] \qquad \textbf{(2.115)}$$

BOUNDARY CONDITIONS

In hydraulic systems, non-pipe elements at each end and each junction must be mathematically modelled as boundary condition. In this present work, the study is limited to one-dimensional problems (straight line); elbow systems require equations of lateral movement of the pipe. The boundary conditions are those given at the reservoir and at the valve. Moreover, anchor conditions also matter since they determine the junction coupling effect.

Hydraulic Configurations

In pipeline systems, fluid transient can be caused by the fast closure of the valve, which can be either upstream or downstream the pipe. Downstream-valve systems (Fig. **2.5**) are usually referred to as the reservoir-pipe-valve systems. In this case, boundary conditions consider the closed valve and the upstream reservoir as described in the first chapter (section 1). The upstream reservoir, however, does not matter in case of upstream valve system (Fig. **2.6**); the sign of the reflected wave depends on the inlet downstream reservoir.

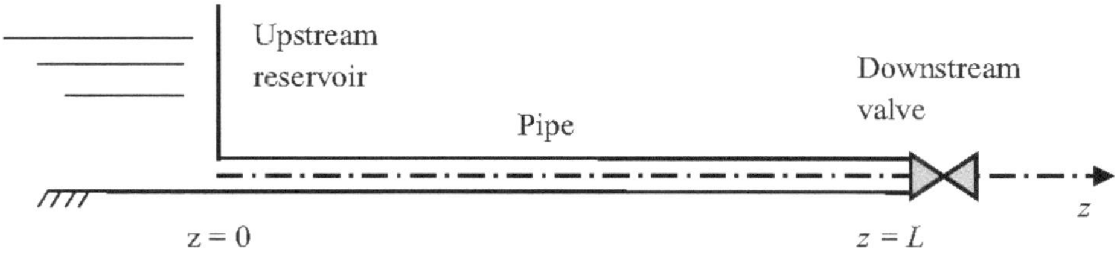

Fig. (2.5). Downstream-valve system (reservoir-pipe-valve-system).

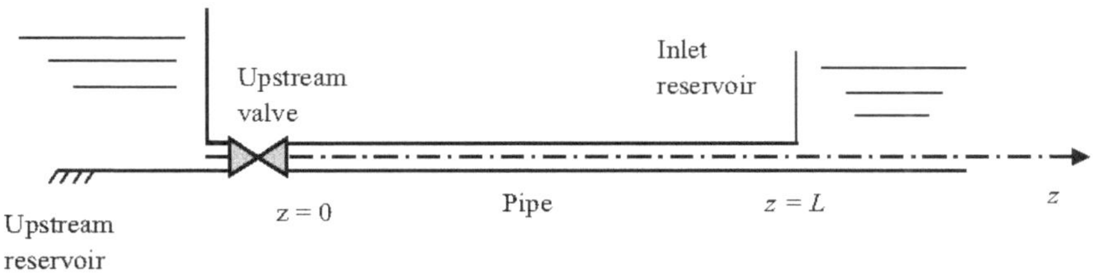

Fig. (2.6). Upstream-valve system.

Reservoirs and Pressurized Tanks

The elevation of the hydraulic grade line (HGL) can be either constant or time-dependent during a short-duration transient. At a large reservoir, the HGL can be normally assumed to be constant, and the following boundary conditions are considered:

$$p(0) = p_{res}\text{, } i.e.\quad H(0) = H_{res} \tag{2.116}$$

$$\dot{u}_z(0) = 0 \tag{2.117}$$

If the reservoir level changes in a known manner, say as a sine wave, the boundary condition is (Wylie and Streeter, 1993):

$$H(0) = H_{res} + \Delta H \sin \omega t \tag{2.118}$$

in which ω is the circular frequency and ΔH is the amplitude of the wave.

Instantaneous Closure Valve

In this subsection, an atmospheric downstream valve is considered (Fig. **2.7**). The dynamic and the static effect of the pipe on the valve are considered; the pressure of the fluid downstream the valve is assumed to be zero (the absolute pressure is equal to barometric pressure) and there is no axial stress. If junction coupling is ignored (cases of rigidly anchored valves), both the fluid velocity and the pipe velocity are equal to zero which implies two boundary conditions. However, if junction coupling is considered for the freely moving valve, the first boundary condition in which neither of the two variables is maintained equal to zero is

$$V_z(L) = \dot{u}_z(L) \qquad \textbf{(2.119)}$$

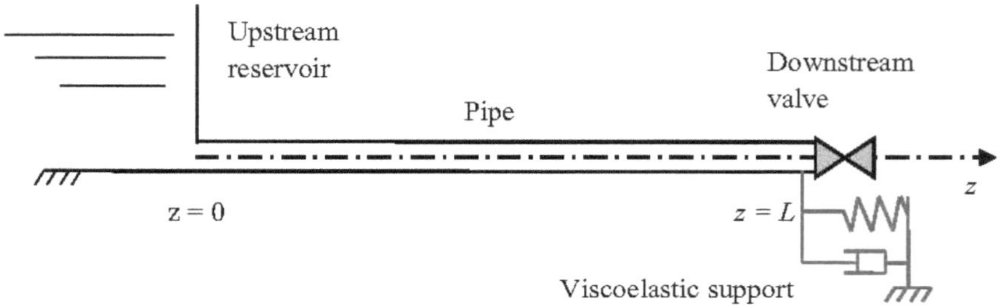

Fig. (2.7). Atmospheric valve attached to viscoelastic support.

In this case, the equation of motion should be used [2, 74, 75]. Obviously, the vibrating downstream valve of mass m is subjected to pressure force, elastic force of the pipe and equivalent stiffness and damping forces generated by the pipe and the support. The equation of motion is given as [2]:

$$m\ddot{u}_z + c\dot{u}_z + ku_z = Ap - A_p\sigma_z \qquad \textbf{(2.120)}$$

with c is the equivalent viscous damping coefficient, k is the stiffness and A and A_p are, respectively, cross sections of the fluid and the pipe. Noting that c and k are in fact equivalent properties of the system composed by the pipeline and the supports [75]. The determination of the damping coefficient of the pipeline can be obtained from the simple oscillator theory as $c = 2\xi A_p\sqrt{E\rho_p}$ [76], with ξ is the damping ratio and E and ρ_p are, respectively, the Young's modulus of elasticity and the density of the pipe-wall. The spring-stiffness of a pipeline of a length L is simply obtained by $k = EA_p/L$.

Non-instantaneous Closure Valve

Unlike the instantaneous closure valve, the non-instantaneous closure valves are treated as an orifice with the following flow rate law:

$$\frac{V_z}{V_{z0}} = \pm\tau(t)\sqrt{\frac{\Delta p}{\Delta p_0}} \tag{2.121}$$

in which Δp_0 is the steady state pressure loss over the valve, Δp is the transient pressure loss caused by the valve closing, V_{z0} is the steady state velocity of the fluid and $\tau(t)$ is a given function of time characterizing the valve opening or closing. In steady state $\tau(t)=1$, however this function is defined with empirical formulae relative to the valve in case of transient flow. The transient function takes account of the critical time of closure T_c and is usually non-linear.

CONCLUSION

In this chapter, the coupled mathematical modelling is developed based on the continuity equation and the momentum equation for both the fluid and the pipe. The modelling takes account of elasticity, unsteady friction (UF), and viscoelasticity. Thus, three coupled models are detailed. The linear elasticity theory (generalized Hooke's law) and the Timoshenko's beams theory are strongly present. Four variables were involved: the pressure and the axial velocity of the fluid and the axial stress and the axial velocity of the pipe. Three types of dynamic coupling are considered: friction coupling, Poisson coupling and junction coupling. The four-equation model (4EM) describes water hammer in elastic pipelines with FSI effect. To better simulate the friction effect, the four-equation friction model (4EFM) is proposed. It is obtained by incorporating an UF model such as the Zielke model and the Vardy-Brown model into the 4EM. The viscoelastic behaviour of the pipe material is also studied in this chapter. The formulation of the classical viscoelastic model (VEM) is detailed, and the four-equation viscoelastic model (4EVEM) is proposed and formulated.

CHAPTER 3

Numerical Calculation of Water Hammer

Abstract: In this chapter, the Method of Characteristics (MOC) is detailed and used to solve the mathematical models defined in the second chapter, where the reservoirpipe-valve-system is considered. The wave-speed adjustment (WSA) scheme is preferred for the MOC. The numerical method is applied to the elastic pipelines on onehand and to the viscoelastic pipelines on the other hand.

Keywords: Boundary conditions; Initial conditions; Linear interpolation; Method of characteristics; Wave-speed adjustment.

INTRODUCTION

The partial derivative equations (PDE) describe numerous physical phenomena such as wave propagation. They can be classified with respect to their order (usually 1st or 2nd order) and their coefficients (linear, quasi-linear or nonlinear). The unknowns usually depend on time t and space x or z. The PDE problems are successfully solved by means of the MOC. However, this method needs numerical schemes and computational grids. Since interpolation leads to numerical oscillations in the result. The WSA scheme is preferred and then applied to the water hammer models already developed, such as the four-equation friction model.

THE METHOD OF CHARACTERISTICS (MOC)

The conventional form of 1st linear PDE systems with constant coefficients is:

$$\mathbf{A}\frac{\partial \mathbf{y}}{\partial t}+\mathbf{B}\frac{\partial \mathbf{y}}{\partial z}=\mathbf{r} \tag{3.1}$$

with **A** and **B** are two matrices, **y** is the vector of unknowns and **r** is the right-hand side vector.

Abdelaziz Ghodhbani, Ezzeddine Haj Taïeb, Mohsen Akrout & Sami Elaoud

Compatibility Equations

The characteristic equation of (3.1) is:

$$\det(\mathbf{B} - \lambda\mathbf{A}) = 0 \tag{3.2}$$

Three situations are possible according to the roots of Eq. (3.2): (i) if all roots are reals number, the system 3.1 is hyperbolic; (ii) if all roots are reals and equal, the system is parabolic and (iii) if all roots are imaginary numbers, the system is elliptic.

The Method of Characteristics (MOC) is preferred for calculation of PDE systems. Eq. (3.1) is equivalent to [2]:

$$\frac{\partial \mathbf{y}}{\partial t} + \mathbf{A}^{-1}\mathbf{B}\frac{\partial \mathbf{y}}{\partial z} = \mathbf{A}^{-1}\mathbf{r} \tag{3.3}$$

According to linear algebra, it is known that for every square matrix with strict real elements, it exists a matrix **S** such that:

$$\mathbf{S}^{-1}\left(\mathbf{A}^{-1}\mathbf{B}\right)\mathbf{S} = \mathbf{D} \tag{3.4}$$

in which the matrix **D** is diagonal and is written in case of four unknowns as

$$\mathbf{D} = \begin{bmatrix} \lambda_1 & 0 & 0 & 0 \\ 0 & \lambda_2 & 0 & 0 \\ 0 & 0 & \lambda_3 & 0 \\ 0 & 0 & 0 & \lambda_4 \end{bmatrix}$$

The eigenvalues λ_i are the characteristic directions. Their calculation can be obtained by solving the following equation:

$$\det\left(\mathbf{A}^{-1}\mathbf{B} - \lambda\mathbf{I}^4\right) = 0 \tag{3.5}$$

which is equivalent to:

$$\det\left[\mathbf{A}^{-1}\left(\mathbf{B}-\lambda\mathbf{A}\right)\right]=0 \tag{3.6}$$

To solve the above equation, one needs a regular matrix **T**:

$$\mathbf{T}=\mathbf{S}^{-1}\mathbf{A}^{-1} \tag{3.7}$$

which implies:

$$\mathbf{D}=\mathbf{TBS} \text{ and } \mathbf{TB}=\mathbf{DS}^{-1} \tag{3.8}$$

Eq. (3.7) leads to:

$$\mathbf{S}^{-1}=\mathbf{TA} \tag{3.9}$$

Hence, the introduction of Eq. (3.9) into Eq. (3.8) leads to:

$$\mathbf{TB}=\mathbf{DTA} \tag{3.10}$$

The multiplication of the system (3.1) by the matrix **T** and the consideration of Eq. (3.10) allows a new formulation of the system:

$$\mathbf{TA}\frac{\partial\mathbf{y}}{\partial t}+\mathbf{DTA}\frac{\partial\mathbf{y}}{\partial z}=\mathbf{Tr} \tag{3.11}$$

By considering the vector $\mathbf{v}=\mathbf{TAy}$, it follows:

$$\frac{\partial\mathbf{v}}{\partial t}+\mathbf{D}\frac{\partial\mathbf{v}}{\partial z}=\mathbf{Tr} \tag{3.12}$$

which is detailed in the following:

$$\frac{\partial v_i}{\partial t}+\lambda_i\frac{\partial v_i}{\partial z}=\left(Tr\right)_i \tag{3.13}$$

The differentiation of v_i gives:

$$\frac{dv_i}{dt}=\frac{\partial v_i}{\partial t}+\frac{dz}{dt}\frac{\partial v_i}{dz} \tag{3.14}$$

which leads to the compatibility equations of the system.

$$\frac{dv_i}{dt} = (Tr)_i \text{, for } \frac{dz}{dt} = \lambda_i \tag{3.15}$$

Numerical Integration Method

The compatibility equations (3.15) are numerically integrated along the characteristic directions λ_1, λ_2, λ_3 and λ_4. The points A_1, A_2, A_3 and A_4 belong respectively to the characteristic lines of slope λ_1, λ_2, λ_3 and λ_4 defined in the (z, t) plan. The numerical integration of the system (3.15) gives:

$$v_i(P) - v_i(A_i) = \int_{A_i}^{P} (Tr)_i \, dt \tag{3.16}$$

The calculation of the right term depends on the term $(Tr)_i$ whether it is a function of time or not. If $(Tr)_i$ is not a time function, then the integration is obtained directly. Otherwise, a numerical method should be used. The Gauss's method is preferred in that case.

$$\int_{A_i}^{P} (Tr)_i \, dt = Tr(A_i)\left(t_P - t_{A_i}\right) \tag{3.17}$$

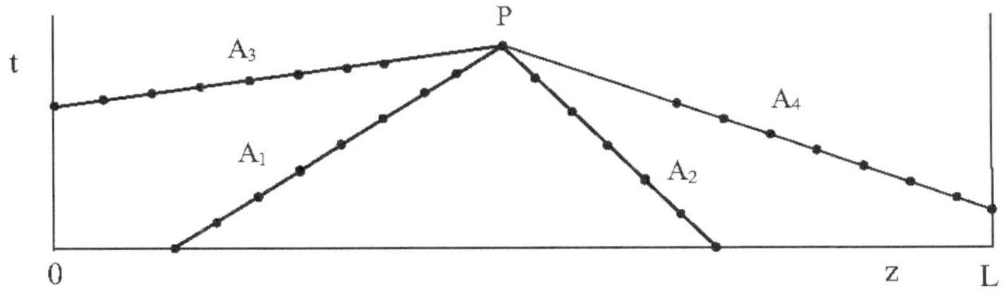

Fig. (3.1). Characteristic direction in the space-time plan.

The numerical integration requires an adequate numerical scheme, such as the space-line interpolation, the time-line interpolation and wave-speed adjustment schemes.

Linear Interpolation

For each unknown y_i, the **space-line interpolation (SLI)** scheme shown in Fig. **(3.2)** allows:

$$\frac{y_i(Q)-y_i(A_1)}{\Delta z'}=\frac{y_i(Q)-y_i(A_3)}{\Delta z} \tag{3.18}$$

which implies:

$$y_i(A_1)=\left(1-\frac{\Delta z'}{\Delta z}\right)y_i(Q)+\frac{\Delta z'}{\Delta z}y_i(A_3) \tag{3.19}$$

And since:

$$\frac{\Delta z'}{\Delta z}=\frac{\Delta z'}{\Delta t}\frac{\Delta t}{\Delta z}=\frac{\lambda_1}{\lambda_3} \tag{3.20}$$

Then:

$$y_i(A_1)=\left(1-\frac{\lambda_1}{\lambda_3}\right)y_i(Q)+\frac{\lambda_1}{\lambda_3}y_i(A_3) \tag{3.21a}$$

And:

$$y_i(A_2)=\left(1-\frac{\lambda_1}{\lambda_3}\right)y_i(Q)+\frac{\lambda_1}{\lambda_3}y_i(A_4) \tag{3.21b}$$

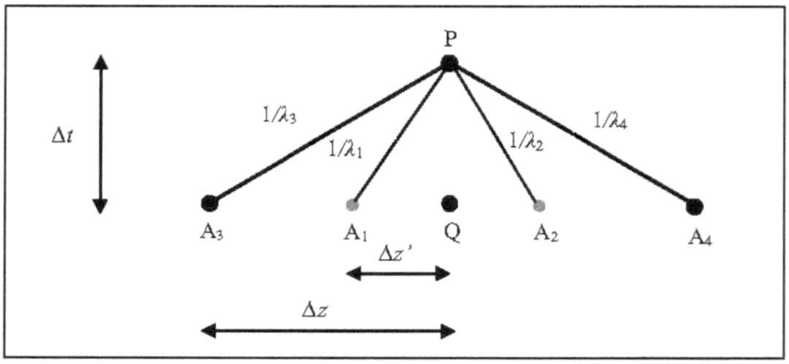

Fig. (3.2). The space-line interpolation (SLI) scheme.

The **time-line interpolation (TLI)** presented in Fig. **(3.3)** takes account of the integer part of the ratio λ_3/λ_1 denoted $In(\lambda_3/\lambda_1)$.

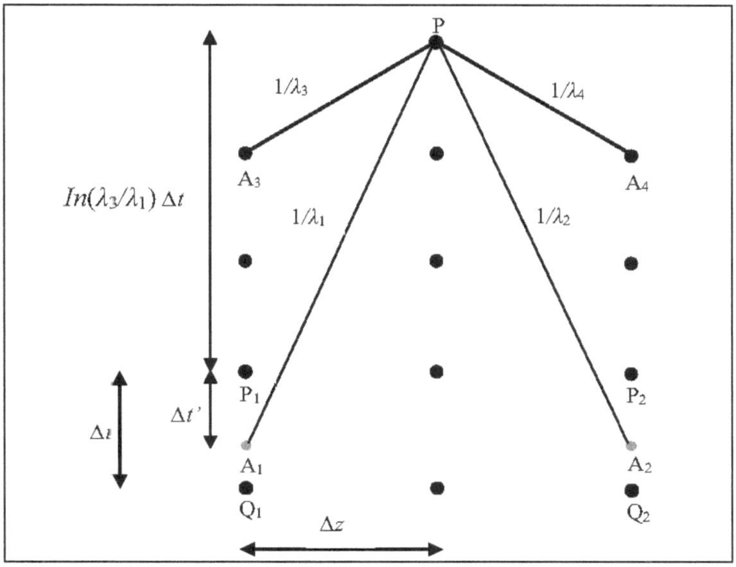

Fig. (3.3). The time-line interpolation (TLI) scheme.

For each unknown y_i, the space-line interpolation (SLI) scheme shown by Fig. **(3.2)** involves:

$$\frac{y_i(P_1)-y_i(Q_1)}{\Delta t}=\frac{y_i(P_1)-y_i(A_1)}{\Delta t'} \tag{3.22a}$$

which results in:

$$y_i(A_1)=\frac{\Delta t'}{\Delta t}y_i(Q_1)+\left(1-\frac{\Delta t'}{\Delta t}\right)y_i(P_1) \tag{3.22b}$$

and since:

$$\frac{In(\lambda_3/\lambda_1)\Delta t+\Delta t'}{\Delta z}=\frac{1}{\lambda_1} \tag{3.23}$$

which is equivalent to:

$$\frac{\Delta t'}{\Delta t} = \frac{\lambda_3}{\lambda_1} - In(\lambda_3/\lambda_1) \tag{3.24}$$

Then:

$$y_i(A_1) = \left(\frac{\lambda_3}{\lambda_1} - In(\lambda_3/\lambda_1)\right) y_i(Q_1) + \left(1 - \frac{\lambda_3}{\lambda_1} + In(\lambda_3/\lambda_1)\right) y_i(P_1) \tag{3.25}$$

And similarly:

$$y_i(A_2) = \left(\frac{\lambda_3}{\lambda_1} - In(\lambda_3/\lambda_1)\right) y_i(Q_2) + \left(1 - \frac{\lambda_3}{\lambda_1} + In(\lambda_3/\lambda_1)\right) y_i(P_2) \tag{3.26}$$

Wave-Speed Adjustment

The **wave-speed adjustment (WSA)** scheme was used in [2, 16, and 77]. The WSA scheme is then used to validate the 4EM and the 4EFM respectively against experimental results [59 and 78]. The approach consists in adjusting the characteristic directions so that all computational points belong the mesh grid. The ratio $n = \lambda_3/\lambda_1$ is modified to be rational. Two integers dented a and b are defined in [59], with $a < b$ so that $n = b/a$. Tijsseling defined two types of computational grid that can be used for the WSA scheme: (i) the pressure-wave grid (Fig. **3.4**) and (ii) the stress-wave grid (Fig. **3.5**) [2].

In the pressure-wave grid, the computation is possible at the first iteration. However, this grid does not allow calculation at boundaries and a refinement is necessary at boundaries. The characteristic directions are chosen as $\lambda_1 = a\Delta z/\Delta t$ and $\lambda_3 = b\Delta z/\Delta t$ and subsequently, the time step Δt is:

$$\Delta t = \frac{a\Delta z}{\lambda_1} = \frac{b\Delta z}{\lambda_3} \tag{3.27}$$

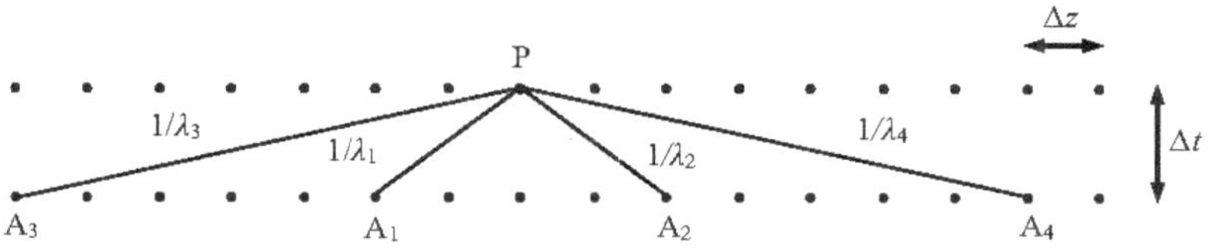

Fig. (3.4). The pressure-wave grid [2].

The stress-wave grid does not require any interpolation at boundaries, but difficulty of calculation at the beginning is involved. The characteristic directions are chosen as $\lambda_1 = \Delta z/(b\Delta t)$ and $\lambda_3 = \Delta z/(a\Delta t)$ and subsequently the time step Δt is:

$$\Delta t = \frac{\Delta z}{b\lambda_1} = \frac{\Delta z}{a\lambda_3} \tag{3.28}$$

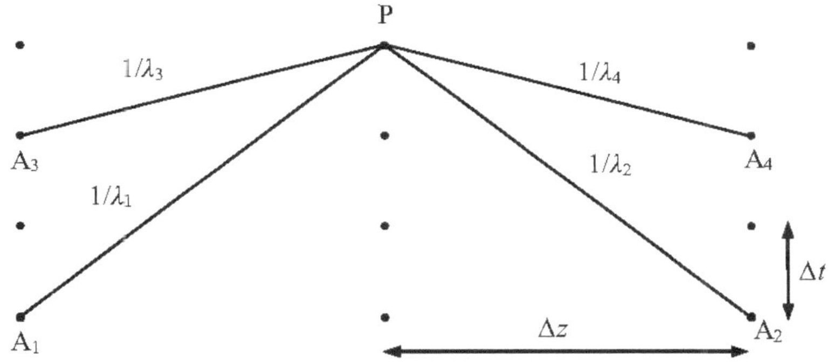

Fig. (3.5). The stress-wave grid [2].

To allow calculation at the beginning using the stress-wave grid, two alternatives are proposed: (i) using SLI for the initial iteration (Fig. **3.6**) and (ii) repeating the steady state conditions for b times until the calculation becomes possible (Fig. **3.7**). The two alternatives lead to the same result since the steady state conditions are linear.

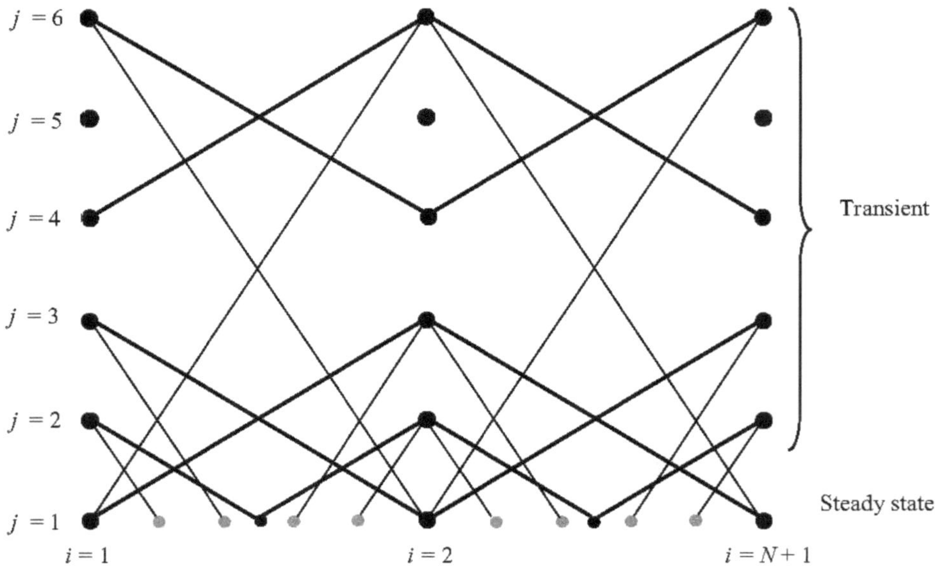

Fig. (3.6). SLI at the initial iteration (grey points) for the stress-wave grid.

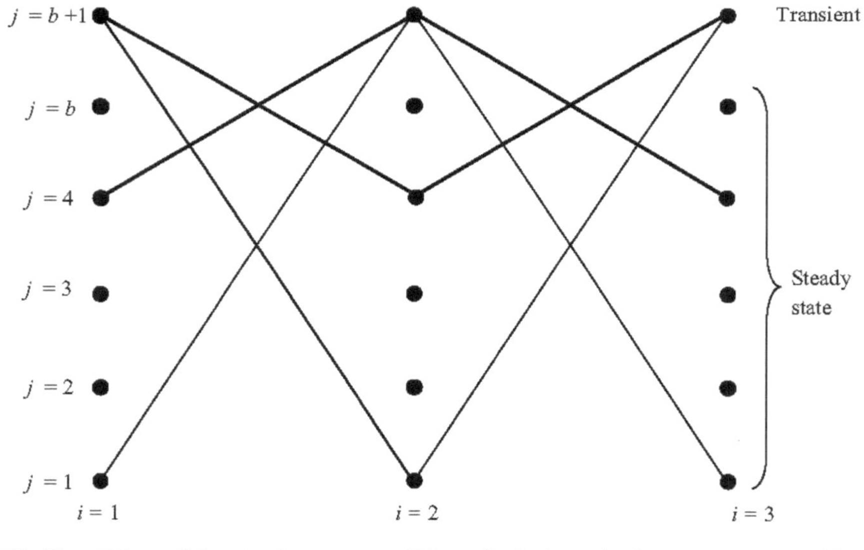

Fig. (3.7). Repetition of the steady state conditions for b times in the stress-wave grid.

WATER HAMMER CALCULATION IN ELASTIC PIPES

The 4EM and the 4EFM which is a generalized form of the former can be calculated in the same manner. Eqs. (2.66) to (2.69) governing the 4EFM can be written in the matrix form described by Eq. (3.1) where the matrices **A**, **B** as well as the unknowns vector **y** and the right-hand side vector **r** are:

$$\mathbf{A}=\begin{bmatrix}1 & 0 & 0 & 0\\ 0 & \dfrac{g}{C_f^2} & 0 & 0\\ 0 & 0 & 1 & 0\\ 0 & \dfrac{\nu D}{2Ee} & 0 & \dfrac{-1}{E}\end{bmatrix};\ \mathbf{B}=\begin{bmatrix}0 & g & 0 & 0\\ 1 & 0 & -2\nu & 0\\ 0 & 0 & 0 & \dfrac{-1}{\rho_p}\\ 0 & 0 & 1 & 0\end{bmatrix};\ \mathbf{y}=\begin{bmatrix}V_z\\ H\\ \dot{u}_z\\ \sigma_z\end{bmatrix};\ \mathbf{r}=\begin{bmatrix}-gh_f\\ 0\\ \dfrac{g\rho_f D}{4e\rho_p}h_f\\ 0\end{bmatrix}$$

Characteristic Directions

Eq. (3.2) can be developed as:

$$\lambda^4-\left(2\rho_f\frac{\nu^2}{E}\frac{R}{E}C_p^2C_f^2+C_p^2+C_f^2\right)\lambda^2+C_p^2C_f^2=0 \quad \textbf{(3.29)}$$

in which the celerity C_f and C_p are defined by Eqs. (1.7) and (1.8), respectively. The calculation of Eq. (3.29) leads to two real roots as:

$$\lambda^2=\frac{1}{2}\left[q^2\mp\left(q^4+4C_p^2C_f^2\right)^{1/2}\right] \quad \textbf{(3.30)}$$

With:

$$q^2=2\nu^2\frac{\rho_f}{\rho_p}\frac{R}{e}C_f^2+C_p^2+C_f^2 \quad \textbf{(3.31)}$$

Consequently, the characteristic directions are derived:

$$\lambda_1=+\frac{1}{\sqrt{2}}\left[q^2-\left(q^4-4C_p^2C_f^2\right)^{1/2}\right]^{1/2}=+\tilde{C}_f=-\lambda_2 \quad \textbf{(3.32)}$$

$$\lambda_3 = +\frac{1}{\sqrt{2}}\left[q^2 + \left(q^4 - 4C_p^2 C_f^2\right)^{1/2}\right]^{1/2} = +\tilde{C}_p = -\lambda_4 \quad \textbf{(3.33)}$$

Determination of the Matrix T

The components of the matrix **T** can be obtained as follows [2]:

$$-\lambda_i T_{i1} + T_{i2} = 0$$

$$T_{i1} - \frac{\lambda_i}{C_f^2} T_{i2} - \frac{\nu R \rho_f}{Ee} \lambda_i T_{i4} = 0 \quad \textbf{(3.34)}$$

$$-2\nu T_{i2} - \lambda_i T_{i3} + T_{i4} = 0$$

$$-C_p^2 T_{i3} + \lambda_i T_{i4} = 0$$

which can be written in matrix form:

$$(\mathbf{B} - \lambda_i \mathbf{A})^{\mathbf{t}}\, \mathbf{T_i} = \mathbf{0} \quad \textbf{(3.35)}$$

To avoid the trivial solution 0 = 0 when $\nu = 0$, the determinant of the above matrices should be equal to zero. Eq. (3.2) which represents the characteristic equation of (3.1) should be considered. Hence, it becomes possible to omit one equation among the system (3.7).

For $i = 1, 2$ *i.e.* $\lambda_i = \pm C_f$, the first and the second equations of the system (3.6) are dependent. By eliminating the second equation and taking $T_{i2} = \lambda_i$, it follows:

$$T_{i1} = 1$$

$$T_{i2} = \lambda_i$$

$$T_{i3} = \frac{2\nu\lambda_i^2}{C_p^2 - \lambda_i^2} \quad \textbf{(3.36)}$$

$$T_{i4} = \frac{2\nu\lambda_i C_p^2}{C_p^2 - \lambda_i^2}$$

For $i = 3, 4$ *i.e.* $\lambda_i = \pm C_p$, the third and the fourth equations of the system (3.6) are dependent. By eliminating the fourth equation and by taking $T_{i4} = \lambda_i$, it follows:

$$
\begin{aligned}
T_{i1} &= \frac{\nu \rho_f R C_f^2 \lambda_i^2}{Ee\left(C_f^2 - \lambda_i^2\right)} \\
T_{i2} &= \frac{\nu \rho_f R C_f^2 \lambda_i^3}{Ee\left(C_f^2 - \lambda_i^2\right)} \qquad \textbf{(3.37)} \\
T_{i3} &= 1 - \frac{2\nu^2 \rho_f R C_f^2 \lambda_i^2}{Ee\left(C_f^2 - \lambda_i^2\right)} \\
T_{i4} &= \lambda_i
\end{aligned}
$$

Thus, the matrix **T** is obtained [2].

$$
\mathbf{T} = \begin{bmatrix}
1 & +\tilde{C}_f & \Sigma_f & \dfrac{C_p^2}{\tilde{C}_f}\Sigma_f \\
1 & -\tilde{C}_f & \Sigma_f & -\dfrac{C_p^2}{\tilde{C}_f}\Sigma_f \\
\Sigma_p & \Sigma_p \tilde{C}_p & 1 - 2\nu\Sigma_p & +\tilde{C}_p \\
\Sigma_p & -\Sigma_p \tilde{C}_p & 1 - 2\nu\Sigma_p & -\tilde{C}_p
\end{bmatrix}
$$

in which the dimensionless quantities Σ_f and Σ_p are:

$$
\Sigma_f = \frac{2\nu C_f^2}{C_p^2 - C_f^2} \quad \text{and} \quad \Sigma_p = \frac{\nu \rho_f D}{2Ee} \frac{C_f^2 \tilde{C}_p^2}{C_f^2 - \tilde{C}_p^2} \qquad \textbf{(3.38)}
$$

Finally, the compatibility equations of the 4EM and the 4EFM read as follows:

$$\frac{dV}{dt} \pm K_f g \frac{dH}{dt} + \Sigma_f \frac{d\dot{u}_z}{dt} \mp \frac{\Sigma_f}{\rho_p \tilde{C}_f} \frac{d\sigma_z}{dt} = \left(\Gamma\Sigma_f - 1\right) g h_f + \Sigma_f g \sin\gamma \quad \textbf{(3.39)}$$

$$\frac{dV}{dt} \pm K_p g \frac{dH}{dt} + \Pi \frac{d\dot{u}_z}{dt} \mp \frac{\tilde{C}_p}{\Sigma_p E} \frac{d\sigma_z}{dt} = \left(\Gamma\Pi - 1\right) g h_f + \Pi g \sin\gamma \quad \textbf{(3.40)}$$

in which the dimensionless numbers are defined as:

$$\Pi = \frac{1}{\Sigma_p} - 2\nu \text{ and } \Gamma = \frac{\rho_f}{\rho_p} \frac{D}{4e} \quad \textbf{(3.41)}$$

and K_f and K_p are constants given by:

$$K_f = \frac{\tilde{C}_f}{C_f^2} + \frac{C_p^2 \Sigma_f \nu D \rho_f}{2\tilde{C}_f E e} \text{ and } K_p = \frac{\tilde{C}_p}{C_f^2} + \frac{\nu R \tilde{C}_p \rho_f}{\Sigma_p E e} \quad \textbf{(3.42)}$$

The expression of the head loss h_f determines the type of the model: if h_f is given by the Darcy-Weisbach formula, the water hammer model is the standard 4EM. If h_f is the sum of steady and unsteady friction terms, then Eqs. (3.39) and (3.40) describe the 4EFM proposed in [59].

Numerical Integration of the Compatibility Equations

In a regular mesh grid with space step Δz and time step Δt, the calculation of the four unknowns of the vector **y** needs the integration of the compatibility equations along the characteristic directions λ_i. This integration is obtained directly for the first part, and by using the Gauss method for the second. The calculation at any point P needs information from previous points of the computational grid. Subsequently, the characteristic lines must meet these previous points.

Calculation at Interior Sections

According to (Fig. **3.1**), the calculation requires four previous points from the computational grid so that the integration of the compatibility equations using the WSA scheme allows for interior points.

$$\left(V_z^P - V_z^{A_1}\right) + K_f g\left(H^P - H^{A_1}\right) + \Sigma_f\left(\dot{u}_z^P - \dot{u}_z^{A_1}\right) - \frac{\Sigma_f}{\rho_p \tilde{C}_f}\left(\sigma_z^P - \sigma_z^{A_1}\right) = b\Delta t g\left[\left(\Gamma\Sigma_f - 1\right)h_f^{A_1} + \Sigma_f \sin\gamma\right] \quad \textbf{(3.43)}$$

$$\left(V_z^P - V_z^{A_2}\right) - K_f g\left(H^P - H^{A_2}\right) + \Sigma_f\left(\dot{u}_z^P - \dot{u}_z^{A_2}\right) + \frac{\Sigma_f}{\rho_p \tilde{C}_f}\left(\sigma_z^P - \sigma_z^{A_2}\right) = b\Delta t g\left[\left(\Gamma\Sigma_f - 1\right)h_f^{A_2} + \Sigma_f \sin\gamma\right] \quad \textbf{(3.44)}$$

$$\left(V_z^P - V_z^{A_3}\right) + K_p g\left(H^P - H^{A_3}\right) + \Pi\left(\dot{u}_z^P - \dot{u}_z^{A_3}\right) - \frac{\tilde{C}_p}{\Sigma_p E}\left(\sigma_z^P - \sigma_z^{A_3}\right) = a\Delta t g\left[\left(\Gamma\Pi - 1\right)h_f^{A_3} + \Pi \sin\gamma\right] \quad \textbf{(3.45)}$$

$$\left(V_z^P - V_z^{A_4}\right) - K_p g\left(H^P - H^{A_4}\right) + \Pi\left(\dot{u}_z^P - \dot{u}_z^{A_4}\right) + \frac{\tilde{C}_p}{\Sigma_p E}\left(\sigma_z^P - \sigma_z^{A_4}\right) = a\Delta t g\left[\left(\Gamma\Pi - 1\right)h_f^{A_4} + \Pi \sin\gamma\right] \quad \textbf{(3.46)}$$

Calculation at Boundaries

The calculation at boundaries consists in considering the previous boundary conditions at the reservoir and at the valve.

At the upstream end (reservoir), the pressure is assumed to be constant whereas the axial displacement of the pipe is neglected. Eqs (3.43) and (3.45) are respectively replaced by:

$$\left(H^P\right)_{z=0} = H_{res} \quad \textbf{(3.47)}$$

$$\left(\dot{u}_z^P\right)_{z=0} = 0 \quad \textbf{(3.48)}$$

At the fixed valve, both fluid and pipe velocities are equal to zero, so that:

$$\left(V_z^P\right)_{z=L} = 0 \quad \textbf{(3.49)}$$

$$\left(\dot{u}_z^P\right)_{z=L} = 0 \quad \textbf{(3.50)}$$

At the freely moving valve of mass m (junction coupling case), the fluid and the pipe act simultaneously. Consequently, the first boundary condition is:

$$\left(V_z^P\right)_{z=L} = \left(\dot{u}_z^P\right)_{z=L} \quad \textbf{(3.51)}$$

The second boundary condition at the valve can be obtained by developing the equation of motion (2.117). Accurate integration can be obtained, thanks to the Newmark method [74]. By taking $\beta = 1/4$ in this method, the differentiation with a backward finite-difference scheme allows:

$$\left(\ddot{u}_z\right)_{t+\Delta t} = \frac{\left(\dot{u}_z\right)_{t+\Delta t} - \left(\dot{u}_z\right)_t}{\Delta t} \quad \textbf{(3.52)}$$

$$\left(u_z\right)_{t+\Delta t} = \left(u_z\right)_t + \frac{\Delta t}{2}\left[\left(\dot{u}_z\right)_{t+\Delta t} + \left(\dot{u}_z\right)_t\right] \quad \textbf{(3.53)}$$

Thus, after incorporating, Eqs. (3.52) and (3.53) in Eq. (2.119), it follows:

$$-A\left(p\right)_{t+\Delta t} + \left(\frac{m}{\Delta t} + c + \frac{k\Delta t}{2}\right)\left(\dot{u}_z\right)_{t+\Delta t} + A_p\left(\sigma_z\right)_{t+\Delta t} = \left(\frac{m}{\Delta t} - \frac{k\Delta t}{2}\right)\left(\dot{u}_z\right)_t - k\left(u_z\right)_t \quad \textbf{(3.54)}$$

WATER HAMMER CALCULATION IN VISCOELASTIC PIPES

Calculation of the Classical Viscoelastic Model

The classical viscoelastic model is described by Eqs. (1.11) and (2.80) which are transformed to the compatibility equations (2.81). The MOC with rectangular computational grid can be used to solve these equations [1]. The integration of the compatibility equations using an implicit finite difference scheme shown by Fig. (C.1) leads to:

$$C_f^{+}: H^P - H^A + \frac{C_f}{gA}\left(q^P - q^A\right) + 2\Delta t \frac{C_f^2}{g}\left(\frac{\partial \varepsilon_{\varphi.r}}{\partial t}\right)^P + \Delta t C_f h_f^A = 0 \quad \textbf{(3.55)}$$

$$C_f^{-}: H^P - H^B - \frac{C_f}{gA}\left(q^P - q^B\right) + 2\Delta t \frac{C_f^2}{g}\left(\frac{\partial \varepsilon_{\varphi.r}}{\partial t}\right)^P - \Delta t C_f h_f^B = 0 \quad \textbf{(3.56)}$$

Development of the Viscoelastic Terms

The integration of the retarded circumferential strain needs some mathematical development.

By assuming constant inner diameter and thickness of the pipe, Eq. (2.77) allows the expression of the retarded circumferential strain as:

$$\left(\varepsilon_{\varphi.r}\right)_t = \frac{D}{2e}\rho_f g \int_0^t \left((H)_{t-s} - H_0\right)\frac{\partial (J)_s}{\partial s} ds \quad \textbf{(3.57)}$$

in which the partial derivative of the creep function can be derived from Eq. (2.78) as:

$$\frac{\partial (J)_s}{\partial s} = \sum_{k=1}^{N} \frac{J_k}{\tau_k} e^{-s/\tau_k} \quad \textbf{(3.58)}$$

This results in:

$$\left(\varepsilon_{\varphi.r}\right)_t = \frac{D}{2e}\rho_f g \int_0^t \left((H)_{t-s} - H_0\right)\sum_{k=1}^{N} \frac{J_k}{\tau_k} e^{-s/\tau_k} ds = \sum_{k=1}^{N}\left(\varepsilon_{\varphi.r}^k\right)_t \quad \textbf{(3.59)}$$

in which the retarded strain for each kelvin-Voight element k is:

$$\left(\varepsilon_{\varphi.r}^k\right)_t = \frac{J_k}{\tau_k}\frac{D}{2e}\rho_f g \int_0^t \left((H)_{t-s} - H_0\right) e^{-s/\tau_k} ds \quad \textbf{(3.60)}$$

And the time derivative of the elementary retarded strain is:

$$\frac{\partial\left(\varepsilon_{\varphi.r}^{k}\right)_t}{\partial t}=\frac{J_k}{\tau_k}\frac{D}{2e}\rho_f g\frac{\partial}{\partial t}\left(\int_0^t\left((H)_{t-s}-H_0\right)e^{-s/\tau_k}ds\right) \tag{3.61}$$

The calculation of the time derivative part in Eq. (3.61) needs to change the integration variable to $u = t - s$, *i.e.* $ds = - du$, it follows [1]:

$$\int_0^t\left((H)_{t-s}-H_0\right)e^{-s/\tau_k}ds=e^{-t/\tau_k}\int_0^t\left((H)_u-H_0\right)e^{u/\tau_k}du \tag{3.62}$$

And:

$$\frac{\partial\left(\varepsilon_{\varphi.r}^{k}\right)_t}{\partial t}=\frac{J_k}{\tau_k}\frac{D}{2e}\rho_f g\left[\begin{array}{l}-\dfrac{e^{-t/\tau_k}}{\tau_k}\displaystyle\int_0^t\left((H)_u-H_0\right)e^{u/\tau_k}du+\\ e^{-t/\tau_k}\dfrac{\partial}{\partial t}\left(\displaystyle\int_0^t\left((H)_u-H_0\right)e^{u/\tau_k}du\right)\end{array}\right] \tag{3.63}$$

The Leibniz integral rule allows:

$$\begin{aligned}\frac{\partial}{\partial t}\left(\int_0^t\left((H)_u-H_0\right)e^{u/\tau_k}du\right)&=\left[\left((H)_u-H_0\right)e^{u/\tau_k}\right]_{u=t}-\left[\left((H)_u-H_0\right)e^{u/\tau_k}\right]_{u=0}\\&=\left((H)_t-H_0\right)e^{t/\tau_k}\end{aligned} \tag{3.64}$$

As a result, the time derivative of the elementary retarded strain becomes:

$$\frac{\partial\left(\varepsilon_{\varphi.r}^{k}\right)_t}{\partial t}=\frac{J_k}{\tau_k}\frac{D}{2e}\rho_f g\left((H)_t-H_0\right)-\frac{1}{\tau_k}\left(\varepsilon_{\varphi.r}^{k}\right)_t \tag{3.65}$$

The next step is the calculation of the elementary retarded strain by incorporating the time step Δt according to the rectangular grid (RG) of the MOC [1].

$$\left(\varepsilon_{\varphi.r}^{k}\right)_t=\frac{J_k}{\tau_k}\frac{D}{2e}\rho_f g\left(\begin{array}{l}\displaystyle\int_0^{t-\Delta t}\left((H)_u-H_0\right)e^{(u-t)/\tau_k}du+\\ \displaystyle\int_{t-\Delta t}^{t}\left((H)_u-H_0\right)e^{(u-t)/\tau_k}du\end{array}\right) \tag{3.66}$$

The one time-step earlier retarded strain can be derived according to Eq. (3.60) as:

$$\left(\varepsilon_{\varphi.r}^{k}\right)_{t-\Delta t}=\frac{J_k}{\tau_k}\frac{D}{2e}\rho_f g\int_0^{t-\Delta t}\left((H)_u-H_0\right)e^{(u-t+\Delta t)/\tau_k}du=$$
$$\frac{J_k}{\tau_k}\frac{D}{2e}\rho_f g e^{\Delta t/\tau_k}\int_0^{t-\Delta t}\left((H)_u-H_0\right)e^{(u-t)/\tau_k}du \quad \textbf{(3.67)}$$

The second term of the right-hand side part of Eq. (3.66) needs the use of the integration by parts rule as:

$$\int_{t-\Delta t}^{t}\left((H)_u-H_0\right)e^{(u-t)/\tau_k}du=$$
$$\left[\left((H)_u-H_0\right)\tau_k e^{(u-t)/\tau_k}\right]_{t-\Delta t}^{t}-\int_{t-\Delta t}^{t}\frac{\partial H}{\partial u}\tau_k e^{(u-t)/\tau_k}du \quad \textbf{(3.68)}$$

in which the first part is:

$$\left[\left((H)_u-H_0\right)\tau_k e^{(u-t)/\tau_k}\right]_{t-\Delta t}^{t}=\tau_k\left[(H)_t-e^{-\Delta t/\tau_k}(H)_{t-\Delta t}+\left(e^{-\Delta t/\tau_k}-1\right)H_0\right] \quad \textbf{(3.69)}$$

while the second part is developed by use of the integration by parts rule in which the second partial derivative of the pressure head H is neglected.

$$\int_{t-\Delta t}^{t}\frac{\partial H}{\partial u}\tau_k e^{(u-t)/\tau_k}du=\left[\frac{\partial H}{\partial u}\tau_k^2 e^{(u-t)/\tau_k}\right]_{t-\Delta t}^{t} \quad \textbf{(3.70)}$$

The calculation can be approximated by differentiating the piezometric head H, at time t with one time-step backward and at time $t-\Delta t$ with one time-step upward in the numerical scheme of (Fig. **C.1**) (rectangular grid).

$$\int_{t-\Delta t}^{t}\frac{\partial H}{\partial u}\tau_k e^{(u-t)/\tau_k}du=\frac{\tau_k^2}{\Delta t}\left(1-e^{-\Delta t/\tau_k}\right)\left((H)_t-(H)_{t-\Delta t}\right) \quad \textbf{(3.71)}$$

Therefore, according to Eqs. (3.66) to (3.71), the elementary retarded strain is obtained as:

$$\left(\varepsilon_{\varphi.r}^{k}\right)_{t} \approx e^{-\Delta t/\tau_k}\left(\varepsilon_{\varphi.r}^{k}\right)_{t-\Delta t} + J_k \frac{D}{2e}\rho_f g \left\{ \begin{array}{l} \left[1-\frac{\tau_k}{\Delta t}\left(1-e^{-\Delta t/\tau_k}\right)\right](H)_t + \left[\frac{\tau_k}{\Delta t}\left(1-e^{-\Delta t/\tau_k}\right)-e^{-\Delta t/\tau_k}\right](H)_{t-\Delta t} \\ -\left(1-e^{-\Delta t/\tau_k}\right)H_0 \end{array} \right\} \quad \textbf{(3.72)}$$

The incorporation of Eq. (3.72) into Eq. (3.65) gives:

$$\frac{\partial\left(\varepsilon_{\varphi.r}^{k}\right)_{t}}{\partial t} \approx -\frac{e^{-\Delta t/\tau_k}}{\tau_k}\left(\varepsilon_{\varphi.r}^{k}\right)_{t-\Delta t} + \frac{J_k}{\Delta t}\frac{D}{2e}\rho_f g\left(1-e^{-\Delta t/\tau_k}\right)(H)_t + \frac{D}{2e}\rho_f g\left\{\left[\frac{J_k}{\tau_k}e^{-\Delta t/\tau_k} - \frac{J_k}{\Delta t}\left(1-e^{-\Delta t/\tau_k}\right)\right](H)_{t-\Delta t} - \frac{J_k}{\tau_k}e^{-\Delta t/\tau_k}H_0\right\} \quad \textbf{(3.73)}$$

The mathematical properties of the retarded strain allow:

$$\frac{\partial\left(\varepsilon_{\varphi.r}\right)_{t}}{\partial t} = \sum_{k=1}^{N}\frac{\partial\left(\varepsilon_{\varphi.r}^{k}\right)_{t}}{\partial t} \quad \textbf{(3.74)}$$

Subsequently, the expression of the retarded circumferential strain rate can be defined as follows:

$$\frac{\partial\left(\varepsilon_{\varphi.r}\right)_{t}}{\partial t} = -\sum_{k=1}^{N}\frac{e^{-\delta\Delta t/\tau_k}}{\tau_k}\left(\varepsilon_{\varphi.r}^{k}\right)_{t-\delta\Delta t} + \frac{D}{2e}\rho_f g\sum_{k=1}^{N}\left(1-e^{-\delta\Delta t/\tau_k}\right)\frac{J_k}{\delta\Delta t}(H)_t + \frac{D}{2e}\rho_f g\sum_{k=1}^{N}\left[\frac{J_k}{\tau_k}e^{-\delta\Delta t/\tau_k} - \frac{J_k}{\delta\Delta t}\left(1-e^{-\delta\Delta t/\tau_k}\right)\right](H)_{t-\delta\Delta t} - \frac{D}{2e}\rho_f g\sum_{k=1}^{N}\frac{J_k}{\tau_k}e^{-\delta\Delta t/\tau_k}H_0 \quad \textbf{(3.75)}$$

where the coefficient δ indicates the type of the computational grid: rectangular grid (RG) or staggered grid (SG), $\delta = 1$ for RG and $\delta = 2$ for SG.

Calculation at Interior Sections

The existence of the unknown head $(H)_t$ in the expression of the retarded strain rate leads to rearrangement of the compatibility equations (3.55) and (3.56) so that the discharge q^P can be easily obtained. However, since the computation using the MOC can be either done in RG or SG, one can use the coefficient δ as mentioned above, and the discharge is:

$$q^P = \frac{1}{\delta B}\left(H^A - H^B\right) + \frac{1}{2}\left(q^A + q^B\right) - \frac{1}{2}\Delta t g A\left(h_f^A + h_f^B\right) \quad \textbf{(3.76)}$$

where $B = C_f/(gA)$ is the pipeline impedance (see Subsection 1.3 in Chapter 4).

According to the rectangular computational grid (Fig. **C.1**), the point P corresponds to computation at section i and time t modelled by (i, j) and the point Q corresponds to computation at section i and time $t - \Delta t$ modelled by $(i, j-1)$. This implies that the subscripts t and $t - \Delta t$ in Eq. (3.75) can be respectively replaced by the superscripts P and Q. If the SG is used, the point Q corresponds rather to $t - \Delta t$. The incorporation of Eq. (3.75) into Eqs. (3.55) and (3.56) leads to the determination of the piezometric head H^P which is valid for both RG and SG according to the coefficient δ ($\delta = 1$ for RG and $\delta = 2$ for SG).

$$\begin{aligned} H^P = {} & 2\beta\Delta t \frac{C_f^2}{g}\sum_{k=1}^{N}\frac{e^{-\delta\Delta t/\tau_k}}{\tau_k}\left(\varepsilon_{\varphi.r}^k\right)^Q + \frac{\beta}{2}\tilde{H}_r + \\ & -\beta\frac{D}{e}\rho_f\Delta t C_f^2\left(\sum_{k=1}^{N}\left[\frac{J_k}{\tau_k}e^{-\delta\Delta t/\tau_k} - \frac{J_k}{\delta\Delta t}\left(1 - e^{-\delta\Delta t/\tau_k}\right)\right]H^Q - \sum_{k=1}^{N}\frac{J_k}{\tau_k}e^{-\delta\Delta t/\tau_k}H_0\right) \end{aligned} \quad \textbf{(3.77)}$$

in which:

$$\tilde{H}_r = H^A + H^B + B\left(q^A - q^B\right) - \Delta t C_f\left(h_f^A - h_f^B\right) \quad \textbf{(3.78)}$$

$$\beta = \left(1 + \frac{D}{e}\rho_f C_f^2 J\right)^{-1} \quad \textbf{(3.79)}$$

With:

$$J = \sum_{k=1}^{N} \left(1 - e^{-\delta \Delta t / \tau_k}\right) \frac{J_k}{\delta} \tag{3.80}$$

Calculation at Boundaries

The calculation at boundaries needs boundary conditions (chapter 2). Two hydraulic configurations are considered: the upstream-valve system and the downstream-valve system.

Reservoirs

The reservoirs (whether with constant head or variable head) impose a known head H_{res} wherever their location (upstream or downstream), whereas the discharge depend on this location. The discharges at upstream reservoirs is defined as:

$$q^P = q^B + \frac{1}{B}\left(H_{res} - H^B\right) - \delta \Delta t A C_f \sum_{k=1}^{N} \frac{e^{-\delta \Delta t / \tau_k}}{\tau_k} \left(\varepsilon_{\varphi.r}^{k}\right)^Q - \Delta t g A h_f^B \tag{3.81a}$$

and, at the downstream reservoir (inlet reservoir), it becomes:

$$q^P = q^A - \frac{1}{B}\left(H_{res} - H^A\right) - \delta \Delta t A C_f \sum_{k=1}^{N} \frac{e^{-\delta \Delta t / \tau_k}}{\tau_k} \left(\varepsilon_{\varphi.r}^{k}\right)^Q - \Delta t g A h_f^A \tag{3.81b}$$

Instantaneous Closure Valves

For instantaneous closure valve, the discharge is equal to zero, *i.e.* $Q = 0$, while the piezometric head can be calculated. The piezometric head at the downstream valve is:

$$\begin{aligned} H^P = {} & 2\beta \Delta t \frac{C_f^2}{g} \sum_{k=1}^{N} \frac{e^{-\delta \Delta t / \tau_k}}{\tau_k} \left(\varepsilon_{\varphi.r}^{k}\right)^Q + \beta\left(H^A + B q^A - \Delta t g A h_f^A\right) + \\ & -\beta \frac{D}{e} \rho_f \Delta t C_f^2 \left(\sum_{k=1}^{N} \left[\frac{J_k}{\tau_k} e^{-\delta \Delta t / \tau_k} - \frac{J_k}{\delta \Delta t}\left(1 - e^{-\delta \Delta t / \tau_k}\right) \right] H^Q - \sum_{k=1}^{N} \frac{J_k}{\tau_k} e^{-\delta \Delta t / \tau_k} H_0 \right) \end{aligned} \tag{3.82a}$$

and at upstream valve, it reads:

$$H^P = 2\beta\Delta t \frac{C_f^2}{g} \sum_{k=1}^{N} \frac{e^{-\delta\Delta t/\tau_k}}{\tau_k} \left(\varepsilon_{\varphi.r}^k\right)^Q + \beta\left(H^B - Bq^B + \Delta t g A h_f^A\right) +$$

$$-\beta \frac{D}{e} \rho_f \Delta t C_f^2 \left(\sum_{k=1}^{N} \left[\frac{J_k}{\tau_k} e^{-\delta\Delta t/\tau_k} - \frac{J_k}{\delta\Delta t}\left(1 - e^{-\delta\Delta t/\tau_k}\right) \right] H^Q - \sum_{k=1}^{N} \frac{J_k}{\tau_k} e^{-\delta\Delta t/\tau_k} H_0 \right) \quad \textbf{(3.82b)}$$

Calculation of the 4EVEM

According to Eqs. (2.83) and (2.84) and the approximated expressions of the hoop and axial stresses (2.76) and (2.96), the retarded strains can be written respectively as:

$$\varepsilon_\varphi = \left(1 - \frac{\nu}{2}\right)\sigma_\varphi * dJ = \underbrace{\left(1 - \frac{\nu}{2}\right)\sigma_\varphi J_0}_{\varepsilon_{\varphi.e}} + \underbrace{\left(1 - \frac{\nu}{2}\right)\frac{D}{2e}\rho_f g \int_0^t \left((H)_{t-s} - H_0\right)\frac{\partial (J)_s}{\partial s} ds}_{\varepsilon_{\varphi.r}} \quad \textbf{(3.83)}$$

$$\varepsilon_z = \left(\frac{1}{2} - \nu\right)\sigma_\varphi * dJ = \underbrace{\left(\frac{1}{2} - \nu\right)\sigma_\varphi J_0}_{\varepsilon_{z.e}} + \underbrace{\left(\frac{1}{2} - \nu\right)\frac{D}{2e}\rho_f g \int_0^t \left((H)_{t-s} - H_0\right)\frac{\partial (J)_s}{\partial s} ds}_{\varepsilon_{z.r}} \quad \textbf{(3.84)}$$

Which lead to the linear strain relationship, used for both viscoelastic thin and thick pipes.

$$\varepsilon_{z.r} = \frac{1 - 2\nu}{2 - \nu} \varepsilon_{\varphi.r} \quad \textbf{(3.85)}$$

The 4EVEM governed by Eqs. (2.97) to (2.100) can be written in matrix form shown by Eq. (3.1). The matrices **A** and **B** as well as the vector of unknowns **y** are the same as the 4EFM, yet the right-hand side vector **r** becomes different due to the viscoelastic terms. In the following expression, the subscript "v" in the vector $\mathbf{r}_v$ refers to the viscoelastic pipe.

$$\mathbf{r_v} = \begin{bmatrix} r_{v.1} \\ r_{v.2} \\ r_{v.3} \\ r_{v.4} \end{bmatrix} = \begin{bmatrix} -gh_f \\ \dfrac{4\left(\nu^2-1\right)}{2-\nu}\dfrac{\partial \varepsilon_{\varphi.r}}{\partial t} \\ g\left(\Gamma h_f + \sin\gamma\right) \\ \dfrac{2\nu-1}{2-\nu}\dfrac{\partial \varepsilon_{\varphi.r}}{\partial t} \end{bmatrix} \quad \textbf{(3.86)}$$

in which the circumferential strain rate is obtained by multiplying the expressions 3.75a for rectangular grid (RG) and 3.75b for staggered (SG) by the coefficient $1-\nu/2$ representing the Poisson coupling effect.

The same approach proposed for the numerical calculation of the 4EFM and the 4EM is applied to the 4EVEM to obtain the following compatibility equations:

$$\frac{dV}{dt} \pm K_f g\frac{dH}{dt} + \Sigma_f \frac{d\dot{u}_z}{dt} \mp \frac{\Sigma_f}{\rho_p \tilde{C}_f}\frac{d\sigma_z}{dt} = r_{v.1} \pm \tilde{C}_f r_{v.2} + \Sigma_f r_{v.3} \pm \frac{C_p^2}{\tilde{C}_f}\Sigma_f r_{v.4} \quad \textbf{(3.87)}$$

$$\frac{dV}{dt} \pm K_p g\frac{dH}{dt} + \Pi\frac{d\dot{u}_z}{dt} \mp \frac{\tilde{C}_p}{\Sigma_p E}\frac{d\sigma_z}{dt} = r_{v.1} \pm \tilde{C}_p r_{v.2} + \Pi r_{v.3} \pm \frac{\tilde{C}_p}{\Sigma_p} r_{v.4} \quad \textbf{(3.88)}$$

Hence, the compatibility equations (3.83) and (3.84) can be detailed by writing the right-hand side parts with respect to the unknown variables at time t. In Eq. (3.83), it yields:

$$r_{v.1} \pm \tilde{C}_f r_{v.2} + \Sigma_f r_{v.3} \pm \frac{C_p^2}{\tilde{C}_f}\Sigma_f r_{v.4} = \left(\Sigma_f \Gamma - 1\right) g h_f + \Sigma_f g \sin\gamma$$
$$\pm \hat{C}_1 \frac{D}{2e}\frac{\rho_f g}{\Delta t} J(H)_t \pm \hat{\gamma}_1 \quad \textbf{(3.89)}$$

With:

$$\hat{C}_1 = 2\left(\nu^2 - 1\right)\tilde{C}_f + \left(\frac{1}{2} - \nu\right)\Sigma_f \frac{C_p^2}{\tilde{C}_f} \tag{3.90}$$

And:

$$\begin{aligned} \hat{\gamma}_1 = {} & \hat{C}_1 \frac{D}{2e}\rho_f g \sum_{k=1}^{N}\left[\frac{J_k}{\tau_k} e^{-\delta\Delta t/\tau_k} - \frac{J_k}{\delta\Delta t}\left(1 - e^{-\delta\Delta t/\tau_k}\right)\right](H)_{t-\delta\Delta t} \\ & -\hat{C}_1 \frac{D}{2e}\rho_f g \sum_{k=1}^{N}\frac{J_k}{\tau_k} e^{-\delta\Delta t/\tau_k} H_0 - \hat{C}_1 \sum_{k=1}^{N}\frac{e^{-\delta\Delta t/\tau_k}}{\tau_k}\left(\varepsilon_{\varphi.r}^k\right)_{t-\delta\Delta t} \end{aligned} \tag{3.91}$$

with $\delta = 1$ for RG and $\delta = 2$ for SG. Similarly, the right-hand side of Eq. (3.88) is:

$$\begin{aligned} r_{v.1} \pm \tilde{C}_p r_{v.2} + \Pi r_{v.3} \pm \frac{\tilde{C}_p}{\Sigma_p} r_{v.4} = {} & (\Pi\Gamma - 1) g h_f \pm \\ & \hat{C}_2 \frac{D}{2e}\frac{\rho_f g}{\Delta t} J(H)_t \pm \hat{\gamma}_2 + \Pi g \sin\gamma \end{aligned} \tag{3.92}$$

With:

$$\hat{C}_2 = 2\left(\nu^2 - 1\right)\tilde{C}_p + \left(\frac{1}{2} - \nu\right)\frac{\tilde{C}_p}{\Sigma_p} \tag{3.93}$$

And:

$$\begin{aligned} \hat{\gamma}_2 = {} & \hat{C}_2 \frac{D}{2e}\rho_f g \sum_{k=1}^{N}\left[\frac{J_k}{\tau_k} e^{-\delta\Delta t/\tau_k} - \frac{J_k}{\delta\Delta t}\left(1 - e^{-\delta\Delta t/\tau_k}\right)\right](H)_{t-\delta\Delta t} \\ & -\hat{C}_2 \frac{D}{2\epsilon}\rho_f g \sum_{k=1}^{N}\frac{J_k}{\tau_k} e^{-\delta\Delta t/\tau_k} H_0 - \hat{C}_2 \sum_{k=1}^{N}\frac{e^{-\delta\Delta t/\tau_k}}{\tau_k}\left(\varepsilon_{\varphi.r}^k\right)_{t-\delta\Delta t} \end{aligned} \tag{3.94}$$

Thus, the compatibility equations become:

$$\frac{dV}{dt} \pm K_f g \frac{dH}{dt} + \Sigma_f \frac{d\dot{u}_z}{dt} \mp \frac{\Sigma_f}{\rho_p \tilde{C}_f} \frac{d\sigma_z}{dt} = \left(\Sigma_f \Gamma - 1\right) g h_f + \Sigma_f g \sin\gamma$$
$$\pm \hat{C}_1 \frac{D}{2e} \frac{\rho_f g}{\Delta t} J(H)_t \pm \hat{\gamma}_1 \quad \textbf{(3.95)}$$

$$\frac{dV}{dt} \pm K_p g \frac{dH}{dt} + \Pi \frac{d\dot{u}_z}{dt} \mp \frac{\tilde{C}_p}{\Sigma_p E} \frac{d\sigma_z}{dt} = \left(\Pi\Gamma - 1\right) g h_f +$$
$$\Pi g \sin\gamma \pm \hat{C}_2 \frac{D}{2e} \frac{\rho_f g}{\Delta t} J(H)_t \pm \hat{\gamma}_2 \quad \textbf{(3.96)}$$

The integration of the compatibility equations (3.95) and (3.96) can be obtained at the computational points P (at time *t*) using the WSA scheme and a direct numerical method as:

$$\left(V_z^P - V_z^{A_1}\right) + K_f g\left(H^P - H^{A_1}\right) + \Sigma_f \left(\dot{u}_z^P - \dot{u}_z^{A_1}\right) - \frac{\Sigma_f}{\rho_p \tilde{C}_f}\left(\sigma_z^P - \sigma_z^{A_1}\right) =$$
$$b\hat{C}_1 \frac{D}{2e} \rho_f g J H^P + b\Delta t \hat{\gamma}_1 + b\Delta t g\left(\Sigma_f \Gamma - 1\right) h_f^{A_1} + b\Delta t \Sigma_f g \sin\gamma \quad \textbf{(3.97)}$$

$$\left(V_z^P - V_z^{A_2}\right) - K_f g\left(H^P - H^{A_2}\right) + \Sigma_f \left(\dot{u}_z^P - \dot{u}_z^{A_2}\right) + \frac{\Sigma_f}{\rho_p \tilde{C}_f}\left(\sigma_z^P - \sigma_z^{A_2}\right) =$$
$$-b\hat{C}_1 \frac{D}{2e} \rho_f g J H^P - b\Delta t \hat{\gamma}_1 + b\Delta t g\left(\Sigma_f \Gamma - 1\right) h_f^{A_2} + b\Delta t \Sigma_f g \sin\gamma \quad \textbf{(3.98)}$$

$$\left(V_z^P - V_z^{A_3}\right) + K_p g\left(H^P - H^{A_3}\right) + \Pi\left(\dot{u}_z^P - \dot{u}_z^{A_3}\right) - \frac{\tilde{C}_p}{\Sigma_p E}\left(\sigma_z^P - \sigma_z^{A_3}\right) =$$
$$a\hat{C}_2 \frac{D}{2e} \rho_f g J H^P + a\Delta t \hat{\gamma}_2 + a\Delta t\left(\Pi\Gamma - 1\right) g h_f^{A_3} + a\Delta t \Pi g \sin\gamma \quad \textbf{(3.99)}$$

$$\left(V_z^P - V_z^{A_4}\right) - K_p g\left(H^P - H^{A_4}\right) + \Pi\left(\dot{u}_z^P - \dot{u}_z^{A_4}\right) + \frac{\tilde{C}_p}{\Sigma_p E}\left(\sigma_z^P - \sigma_z^{A_4}\right) =$$
$$-a\hat{C}_2 \frac{D}{2e}\rho_f gJH^P - a\Delta t\hat{\gamma}_2 + a\Delta t\left(\Pi\Gamma - 1\right) gh_f^{A_4} + a\Delta t\Pi g \sin\gamma \quad \textbf{(3.100)}$$

A constant inverse solution matrix can be used to rearrange the above equations is presented in appendix D.

CONCLUSION

The MOC has been used to solve the water hammer models in both elastic and viscoelastic pipelines. The hydraulic configuration used for boundary conditions has considered the reservoir-pipe-valve system defined in the second chapter. The upstream-valve system case can be studied similarly. The WSA scheme and the linear interpolations schemes (SLI and TLI) have been detailed. The MOC development used for elastic pipelines (4EM and the 4EFM) has been extended to the viscoelastic pipelines, for which the 4EVEM has been developed. This latter needed further development because of the existence of the retarded strain in the governing equations.

CHAPTER 4

Numerical Modelling of Transient Cavitation

Abstract: The modelling of transient cavitation is described in this chapter by considering column separation modelling where FSI and gas release are taken into account. As detailed for water hammer modelling, transient in straight pipeline is caused by fast closure of shut-off valve. Both vaporous and gaseous cavitation are considered in the numerical models.

Keywords: Cavity; Collapse; Column separation; Discrete model; Gas release; pressure pulse, Vaporous cavitation.

INTRODUCTION

Transient cavitation occurs in piping systems through two modes: liquid column separation and distributed cavitation. This latter is described in the literature through the bubble model, while the discrete vapour cavity model is widely used when column separation is assumed to occur, which is the case of the present study. In this study, the emphasis is placed on the effect of fluid-structure interaction (FSI) on the solution. In fact, the chapter focuses on the numerical modelling of cavitation by considering four-equation models.

COLUMN SEPARATION MODELLING IN ELASTIC PIPES

The Coupled DVCM

General Concept

The coupled DVCM was proposed [52] to take account of FSI in the vaporous cavitation modelling; the model ignore the gas release effect. Transient is caused by a fast valve closure. Initial velocity is assumed to be high, so that vaporous cavitation and column separation occur. Moreover, thermodynamic conditions characterizing the classical DVCM, such as cavity opening and collapse and the constant vapour pressure in the cavity, are still valid. As described for the classical DVCM, vapour cavities are concentrated at the grid points, whereas liquid is

Abdelaziz Ghodhbani, Ezzeddine Haj Taïeb, Mohsen Akrout & Sami Elaoud overlying the reaches. Nevertheless, the isothermal process is no longer adopted; the cavitation process is assumed to be isentropic.

The coupled DVCM is obtained by incorporating the classical DVCM into the compatibility equations (3.39) and (3.40) deriving from the 4EFM. The MOC based on the WSA scheme is used to solve the proposed model. The space-time plan is divided into regular mesh with a space step Δz and a time step Δt. While $H > H_v - z \sin \gamma$ with γ is the inclination of the pipe, cavitation does not occur, and the water hammer model is used. Once the pressure p drops below the gauge vapour pressure p_v at the computation point P, the strong condition $H = H_v - z \sin \gamma$ is established. Four unknowns are then calculated at each time step Δt: the upstream discharge q_u^P, the downstream discharge q_d^P, the axial velocity of the pipe $\dot{u}_z^P$ and the axial stress σ_z^P.

The coupled DVCM was calculated in a "stress wave" grid used in a specific form yielding two different schemes [52]. The first one is a full-WSA scheme that allows calculations at all points of the RG as shown in Fig. (**4.1**). This scheme can be used whether the integers a and b characterizing the WSA method are odd or even. The second one called WSA-TLI scheme allows calculations with the WSA method only at the first SG, whereas TLI provides information at points of the second SG. This scheme is illustrated in Fig. (**4.1**), where the integers a and b are indicative for calculation only. If a single SG is used instead of the RG, then calculations based on the WSA method cannot be performed unless both integers a and b are odd. Since these latter are chosen either odd or even with respect to the characteristic direction ratio r, the single SG is not useful. Therefore, the WSA-TLI scheme is assumed to be a helpful alternative for the full-WSA because it allows calculation of the coupled DVCM in a SG using the WSA method.

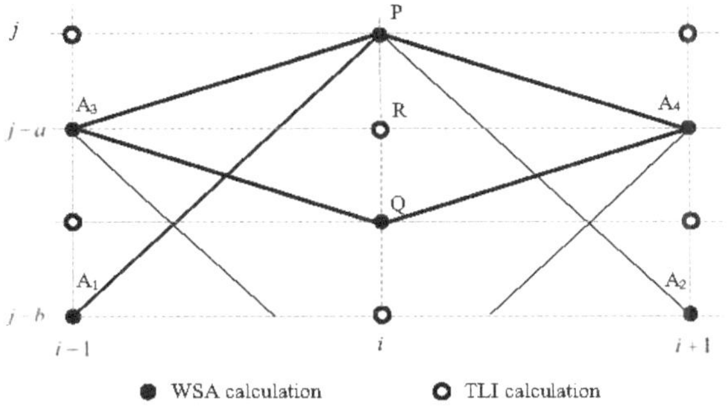

Fig. (4.1). "Stress wave" grid based on the WSA-TLI scheme [52].

The calculation of the hollow points shown in Fig. (**4.1**) using the TLI technique is simply obtained since the time step Δt is constant. For example, the discharge at the point R (earlier point of P) is:

$$q_u^R = \frac{1}{2}\left(q_u^Q + q_u^P\right) \text{ and } q_d^R = \frac{1}{2}\left(q_d^Q + q_d^P\right) \tag{4.1}$$

and so are $\dot{u}_z^R$ and σ_z^R. Then, the cavity volume $\forall_v$ is calculated at each point of the RG and the formula given by Eq. (1.19) is modified to:

$$\forall_v^P = \forall_v^Q + \Delta t\left[\psi\left(q_d^P - q_u^P\right) + (1-\psi)\left(q_d^Q - q_u^Q\right)\right] \tag{4.2}$$

Once the compatibility equations are integrated along the characteristic directions $+\tilde{C}_f$, $-\tilde{C}_f$ $+\tilde{C}_p$ and $-\tilde{C}_p$, the four governing equations of the vaporous cavitating flow are derived:

$$\frac{1}{A}\left(q_u^P - q_d^{A_1}\right) + K_f g\left(H^P - H^{A_1}\right) + \Sigma_f\left(\dot{u}_z^P - \dot{u}_z^{A_1}\right) - \frac{\Sigma_f}{\rho_p \tilde{C}_f}\left(\sigma_z^P - \sigma_z^{A_1}\right) = b\Delta t g\left[\left(\Gamma\Sigma_f - 1\right)h_{fd}^{A_1} + \Sigma_f \sin\gamma\right] \tag{4.3}$$

$$\frac{1}{A}\left(q_d^P - q_u^{A_2}\right) - K_f g\left(H^P - H^{A_2}\right) + \Sigma_f\left(\dot{u}_z^P - \dot{u}_z^{A_2}\right) + \frac{\Sigma_f}{\rho_p \tilde{C}_f}\left(\sigma_z^P - \sigma_z^{A_2}\right) = b\Delta t g\left[\left(\Gamma\Sigma_f - 1\right)h_{fu}^{A_2} + \Sigma_f \sin\gamma\right] \tag{4.4}$$

$$\frac{1}{A}\left(q_u^P - q_d^{A_3}\right) + K_p g\left(H^P - H^{A_3}\right) + \Pi\left(\dot{u}_z^P - \dot{u}_z^{A_3}\right) - \frac{\tilde{C}_p}{\Sigma_p E}\left(\sigma_z^P - \sigma_z^{A_3}\right) = a\Delta t g\left[\left(\Gamma\Pi - 1\right)h_{fd}^{A_3} + \Pi \sin\gamma\right] \tag{4.5}$$

$$\frac{1}{A}\left(q_d^P - q_u^{A_4}\right) - K_p g\left(H^P - H^{A_4}\right) + \Pi\left(\dot{u}_z^P - \dot{u}_z^{A_4}\right) + \frac{\tilde{C}_p}{\Sigma_p E}\left(\sigma_z^P - \sigma_z^{A_4}\right) = a\Delta t g\left[\left(\Gamma\Pi - 1\right) h_{fu}^{A_4} + \Pi \sin\gamma\right] \quad \textbf{(4.6)}$$

with $h_{fu}^{A_i}$ and $h_{fd}^{A_i}$ refer, respectively, to upstream and downstream friction term at a point A_i with respect to upstream and downstream discharge at the same point according to Eq. (E4) where the velocity V_z is replaced by the upstream relative velocity $V_{rel.u} = q_u^{A_i}/A - \dot{u}_z^{A_i}$, respectively, the downstream relative velocity $V_{rel.d} = q_d^{A_i}/A - \dot{u}_z^{A_i}$.

Initial conditions are deduced from steady state conditions as for the 4EM and the 4EFM. However, boundary conditions depend on the hydraulic configuration. At the reservoirs, it is assumed that cavitation does not occur. At a downstream valve for instance, the first boundary condition consists in assigning zero to the downstream discharge.

$$\left(q_d\right)_{z=L} = 0 \quad \textbf{(4.7)}$$

whereas the anchoring condition of the pipeline (rigidly restrained valve) allows the second boundary condition exactly like in case of the 4EM and the 4EFM as $\left(\dot{u}_z\right)_{z=L} = 0$. These two equations replace, respectively, Eqs. (4.4) and (4.6).

The Variable Wave-Speed Method

Besides the use of a SG for the MOC with WSA in calculation, the pressure wave-speed was considered in two different ways [52]. The first one is the standard procedure which supposes that the pressure wave-speed remains constant during the transient event. This assumption has been already adopted in the classical DVCM. The second one, referred to as *variable wave-speeds* (VWS) method consists in modifying the pressure wave-speed across the cavitation region. The DVCM supposes that liquid underlies between nodes whether vapour cavity exists or not in adjacent nodes. Thus, the VWS method considers the existence of a liquid-vapour interface (Fig. **4.2**). Besides, the cavity evolution is assumed to be isentropic unlike the isothermal cavitation considered in the classical DVCM. The isentropic assumption is used in the literature in calculation of sound speed in diphasic flow where barotropic fluids (vapour and liquid) are studied [41]. Nonetheless, the determination of a rigorous formula for the pressure wave-speed in two-phase flow becomes more complicated when cavitation occurs in liquid-fully pipelines. In addition, a pressure dependent wave-speed leads to nonlinearity of the hyperbolic system. Schmidt *et al.* (1999) gave a numerical model that treats liquid and vapour as a continuum for predicting small-scale, high-speed, cavitating nozzle flow. The Schmidt's model considers that cavitating flow is a homogeneous and barotropic mixture of gas and liquid. Moreover, the density ρ of the mixture depends only on the pressure p and subsequently the sound speed C_0 satisfies the following relationship [18]:

$$dp = C_0^2 d\rho \tag{4.8}$$

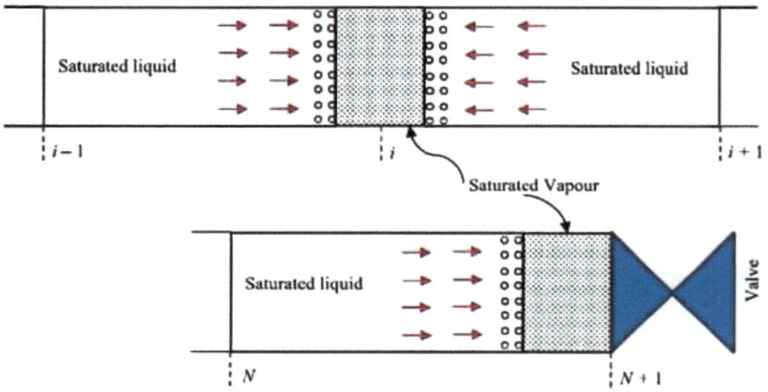

Fig. (4.2). Heat transfer (Red arrows) during cavity growing at interior point (top) and at the valve (bottom) [52].

Denoting $\forall_g$ the gas (vapour) volume in a mixture volume $\forall$ and $\alpha = \forall_g / \forall$ the void fraction, the density of the mixture is:

$$\rho = \alpha \rho_g + (1-\alpha)\rho_l \tag{4.9}$$

with ρ_g and ρ_l are, respectively, densities of saturated vapour and saturated liquid at the pressure p.

Theoretically, the sound speed C in the mixture can be given as [18]:

$$C_0 = \left[\rho \left(\frac{\alpha}{\rho_g C_g^2} + \frac{1-\alpha}{\rho_l C_l^2} \right) \right]^{-1/2} \quad \textbf{(4.10)}$$

with C_g and C_l are, respectively, sound speeds of saturated vapour and saturated liquid. The interface assumption considered in this study implies the existence of a very small void fraction in the liquid before vapour coalescence. As a result, the density, and the sound speed of liquid in the vicinity of the cavity are approximated as the following:

$$\rho \approx \rho_l \text{ and } C_0 \approx C_l \quad \textbf{(4.11)}$$

The introduction of the above approximations in the Korteweg's expression leads to the following simplified formula for the pressure wave-speed in the vicinity liquid:

$$C_f = \left\{ \rho_l \left[\frac{1}{K} + \frac{D}{Ee} \right] \right\}^{-1/2} \quad \textbf{(4.12)}$$

In the governing equations of all coupled models (the 4EM, the 4EFM and 4EVM), it follows:

$$C_f = \left\{ \rho_l \left[\frac{1}{K} + \left(1-\nu^2\right) \frac{D}{Ee} \right] \right\}^{-1/2} \quad \textbf{(4.13)}$$

Regarding the value of ρ_l, the occurrence of column separation at a section i of the pipeline begets a slight decrease in temperature of the vicinity liquid, which simultaneously causes increasing of ρ_l, decreasing in K and subsequently decreasing in the characteristic directions $\tilde{C}_f$ and $\tilde{C}_p$. This phenomenon can be explained by the fact that vaporization of any mass of liquid requires heat to be absorbed from outside of this liquid. Accordingly, column separation involves heat transfer from the vicinity liquid to the cavity across the liquid-vapour interface (**Fig. 4.2**). This heat transfer yields decreasing in temperature of the liquid and then changes its physical properties. More precisely, the isentropic pressure drop of a liquid volume from an initial pressure p_1 and a temperature T_1 to a first saturation state pressure (p_{v1}, T_1) causes vaporization of a portion of this liquid thanks to the transferred heat of the vicinity liquid. The two fluids in contact reach a final saturation state (p_{v2}, T_2). Schmidt (1999) imposed limitation for this process and gives a formula allowing the calculation of the minimum pressure $p_{\min}$. Schmidt (1999) and Liu *et al.* (2004) demonstrated that to get reasonable values of the pressure p, the condition $\rho_l/\rho_g > 10^5$ should be satisfied.

To simplify the algorithm, the following assumptions which imply restriction of the number of reaches N were used in [52]:

(i) The cavity region is limited by Δz at the valve and by $2\Delta z$ at the interior points (Fig. **4.2**), so that the number of reaches N should be as limited as the cavity size remains very small compared to the reach volume. In another word, the cavity length l_{cav} is very small compared to Δz, *i.e.*, $l_{cav}/\Delta z \ll 1$. The classical DVCM can give accurate results, if the ratio of the maximum cavity size to reach volume stay below 10 % [79].

(ii) Although the vapour pressure in the cavity zone is variable according to the isentropic assumption, this variation is ignored in the algorithm. Moreover, for each cavitating section i, computations at sections $i-1$ and $i+1$ do not consider the VWS method, unless column separation occurs at these sections. This restriction can be explained by the small cavity size and the short duration of the cavity existence.

The Coupled DGCM

This subsection investigates the application of the coupled approach, already proposed for the DVCM, to the DGCM. The coupled DGCM model is obtained by incorporating the classical DGCM into the 4EFM. UF is used to reduce oscillations as expected in [52].

Calculation at Interior Sections

As shown in the above equations, the piezometric head H^P at the interior computational point P is an unknown in the system and an expression should be provided for it as given in Eq. (1.42) for the classical DGCM. To obtain such formula, Eqs (4.3) to (4.6) can be rearranged in such a manner the quantity $q_d^P - q_u^P$ is written with respect to H^P. The omission of the unknown $\dot{u}_z^P$ leads to:

$$\frac{q_d^P - q_u^P}{gA} = 2K_f H^P - \frac{2\Sigma_f}{g\rho_p \tilde{C}_f}\sigma_z^P + F_f \tag{4.14}$$

And:

$$\frac{q_d^P - q_u^P}{gA} = 2K_p H^P - \frac{2\tilde{C}_p}{\Sigma_p g E}\sigma_z^P + F_p \tag{4.15}$$

With:

$$F_f = \frac{1}{gA}\left(q_u^{A_2} - q_d^{A_1}\right) - K_f\left(H^{A_2} + H^{A_1}\right) + \frac{\Sigma_f}{g}\left(\dot{u}_z^{A_2} - \dot{u}_z^{A_1}\right) + \frac{\Sigma_f}{g\rho_p \tilde{C}_f}\left(\sigma_z^{A_2} + \sigma_z^{A_1}\right) + b\Delta t\left(\Gamma\Sigma_f - 1\right)\left(h_{fu}^{A_2} - h_{fd}^{A_1}\right) \tag{4.16}$$

And:

$$F_p = \frac{q_u^{A_4} - q_d^{A_3}}{gA} - K_p\left(H^{A_4} + H^{A_3}\right) + \frac{\Pi}{g}\left(\dot{u}_z^{A_4} - \dot{u}_z^{A_3}\right) + \frac{\tilde{C}_p}{g\Sigma_p E}\left(\sigma_z^{A_4} + \sigma_z^{A_3}\right) + a\Delta t\left(\Gamma\Pi - 1\right)\left(h_{fu}^{A_4} - h_{fd}^{A_3}\right) \tag{4.17}$$

Subsequently, Eqs. (4.14) and (4.15) allow:

$$q_d^P - q_u^P = gA\left(2\tilde{K}H^p + \tilde{F}\right) \tag{4.18}$$

where $\tilde{K}$ and $\tilde{F}$ are given by:

$$\tilde{K} = K_f - \frac{\Sigma_f \Phi}{\rho_p \tilde{C}_f}\left(K_p - K_f\right) \text{ and } \tilde{F} = F_f - \frac{\Sigma_f \Phi}{\rho_p \tilde{C}_f}\left(F_p - F_f\right) \tag{4.19}$$

in which:

$$\Phi = \left(\frac{\tilde{C}_p}{\Sigma_p E} - \frac{\Sigma_f}{\rho_p \tilde{C}_f} \right)^{-1} \tag{4.20}$$

Thus, after replacing Eq. (4.18) in Eq. (1.41) and then taking account of Eq. (1.27), the following standard quadratic equation carrying out the variable H^P is obtained:

$$\left(H^P\right)^2 + \left[\tilde{Z} - \left(Z + H_v\right)\right] H^P - \tilde{Z}\left(Z + H_v\right) - \frac{p_0^* \alpha_0 \forall_m}{\rho_f g \tilde{A}} = 0 \tag{4.21}$$

With:

$$\tilde{Z} = \frac{\tilde{F}}{2\tilde{K}} + \frac{1}{\tilde{A}} \left[\forall_g^Q + 2\Delta t \left(1 - \psi\right) \left(q_d^Q - q_u^Q \right) \right] \tag{4.22}$$

And:

$$\tilde{A} = 4 \Delta t \psi g A \tilde{K} \tag{4.23}$$

The resolution of Eq. (4.21) leads in fact to two different real solutions.

$$H_{1,2}^P = \frac{1}{2}\left(Z + H_v - \tilde{Z}\right) \mp \left[\frac{1}{4}\left(\tilde{Z} - Z - H_v\right)^2 + \tilde{Z}\left(Z + H_v\right) + \frac{p_0^* \alpha_0 \forall_m}{\tilde{A} \rho_f g} \right]^{1/2} \tag{4.24}$$

Calculation at the Fixed Valve

In case of downstream-valve piping systems, gaseous cavitation and vaporous cavitation are assumed to not occur at the reservoir. At the downstream valve, column separation can take place, and Eq. (4.4) are replaced by $q_d^P = 0$ whereas Eq. (4.6) are replaced by the expression of the axial velocity according to the anchor conditions. If the valve is assumed to not move, then the condition $\dot{u}_z^P = 0$ is established. A similar procedure allows the expression of the upstream discharge q_u^P with respect of H^P as:

$$q_u^P = gA\left(\tilde{K}_L H^P + \tilde{F}_L\right) \tag{4.25}$$

With:

$$\tilde{K}_L = \frac{\Sigma_f \Phi\left(K_p - K_f\right)}{\rho_p \tilde{C}_f} - K_f \tag{4.26}$$

$$\tilde{F}_L = \frac{q_d^{A_1}}{gA} + K_f H^{A_1} + \frac{\Sigma_f}{g}\dot{u}_z^{A_1} + \frac{\Sigma_f}{g\rho_p \tilde{C}_f}\left(\Phi G - \sigma_z^{A_1}\right) + b\Delta t\left[\left(\Gamma\Sigma_f - 1\right)h_{fd}^{A_1} + \Sigma_f \sin\gamma\right] \tag{4.27}$$

And:

$$G = \frac{q_d^{A_1} - q_d^{A_3}}{A} + g\left(K_f H^{A_1} - K_p H^{A_3}\right) + \Sigma_f \dot{u}_z^{A_1} - \Pi \dot{u}_z^{A_3} + \frac{\tilde{C}_p \sigma_z^{A_3}}{\Sigma_p E} - \frac{\Sigma_f \sigma_z^{A_1}}{\rho_p \tilde{C}_f} + \Delta tg\left[b\left(\Gamma\Sigma_f - 1\right)h_{fd}^{A_1} - a\left(\Gamma\Pi - 1\right)h_{fd}^{A_3} + \left(b\Sigma_f - a\Pi\right)\sin\gamma\right] \tag{4.28}$$

Consequently, the following quadratic equation can be derived.

$$\left(H^P\right)^2 + \left[\tilde{Z}_L - Z_L - H_v\right]H^P - \left(Z_L + H_v\right)\tilde{Z}_L + \frac{p_0^* \alpha_0 \forall_m}{\tilde{A}_L \rho g} = 0 \tag{4.29}$$

With:

$$\tilde{Z}_L = \frac{\tilde{F}_L}{\tilde{K}_L} - \frac{1}{\tilde{A}_L}\left[\forall_g^Q - 2\Delta t\left(1-\psi\right)q_u^Q\right] \tag{4.30}$$

And:

$$\tilde{A}_L = 2\Delta t \psi g A \tilde{K}_L \tag{4.31}$$

The two real solutions are:

$$H_{1,2}^P = \frac{1}{2}\left(Z_L + H_v - \tilde{Z}_L\right) \mp \left[\frac{1}{4}\left(\tilde{Z}_L - Z_L - H_v\right)^2 + \tilde{Z}_L\left(Z_L + H_v\right) - \frac{\overset{*}{p_0}\alpha_0 \forall_m}{\tilde{A}_L \rho g}\right]^{1/2} \quad \textbf{(4.32)}$$

Calculation at the Freely Moving Valve

If the valve can move in axial direction, the axial velocity is no longer assigned to zero in Eqs. (4.8) (4.10) which are respectively transformed to

$$\begin{aligned} q_u^P + \Sigma_f A \dot{u}_z^P - \frac{\Sigma_f A}{\rho_p \tilde{C}_f}\sigma_z^P &= -K_f g A H^P + K_f g A H^{A_1} + \Sigma_f A \dot{u}_z^{A_1} + \\ & q_d^{A_1} - \frac{\Sigma_f A}{\rho_p \tilde{C}_f}\sigma_z^{A_1} + b\Delta t g A\left[\left(\Gamma \Sigma_f - 1\right) h_{fd}^{A_1} + \Sigma_f \sin\gamma\right] \end{aligned} \quad \textbf{(4.33)}$$

And:

$$\begin{aligned} q_u^P + \Pi A \dot{u}_z^P - \frac{\tilde{C}_p A}{\Sigma_p E}\sigma_z^P &= -K_p g A H^P + K_p g A H^{A_3} + \Pi A \dot{u}_z^{A_3} + \\ & q_d^{A_3} - \frac{\tilde{C}_p A}{\Sigma_p E}\sigma_z^{A_3} + a\Delta t g A\left[\left(\Gamma \Pi - 1\right) h_{fd}^{A_3} + \Pi \sin\gamma\right] \end{aligned} \quad \textbf{(4.34)}$$

Moreover, Eq. (4.6) is replaced by Eq. (3.25) which can be written as:

$$\left(\frac{m}{\Delta t} + c + \frac{k\Delta t}{2}\right)\dot{u}_z^P + A_p \sigma_z^P = A\rho_f g H^P + \left(\frac{m}{\Delta t} - \frac{k\Delta t}{2}\right)\dot{u}_z^Q - k u_z^Q + A\rho_f g Z \sin\gamma \quad \textbf{(4.35)}$$

Then, the combination of Eqs. (4.33) to (4.35) allows to write the upstream discharge at the computational point P with respect of the piezometric head at the same point. At the first step, the axial velocity can be omitted from the above equations and this operation results in:

$$\left(\frac{1}{\Sigma_f} - \frac{1}{\Pi}\right) q_u^P = \tilde{\alpha}_1 H^P + \tilde{\beta}_1 \sigma_z^P + \tilde{q}_1 \quad \textbf{(4.36)}$$

$$\frac{1}{\Pi} q_u^P = \tilde{\alpha}_2 H^P + \tilde{\beta}_2 \sigma_z^P + \tilde{q}_2 \quad \textbf{(4.37)}$$

With:

$$\tilde{\alpha}_1 = gA\left(\frac{K_p}{\Pi} - \frac{K_f}{\Sigma_f}\right) \quad \textbf{(4.38)}$$

$$\tilde{\beta}_1 = A\left(\frac{1}{\rho_p \tilde{C}_f} - \frac{\tilde{C}_p}{\Sigma_p E}\right) \quad \textbf{(4.39)}$$

$$\tilde{q}_1 = gA\left(\frac{K_f H^{A_1}}{\Sigma_f} - \frac{K_p H^{A_3}}{\Pi}\right) + A\left(\dot{u}_z^{A_1} - \dot{u}_z^{A_3} + \frac{\tilde{C}_p \sigma_z^{A_3}}{\Pi \Sigma_p E} - \frac{\sigma_z^{A_1}}{\rho_p \tilde{C}_f}\right) + \quad \textbf{(4.40)}$$

$$\frac{q_d^{A_1}}{\Sigma_f} - \frac{q_d^{A_3}}{\Pi} + \Delta t g A\left[b\left(\Gamma - \frac{1}{\Sigma_f}\right) h_{fd}^{A_1} - a\left(\Gamma - \frac{1}{\Pi}\right) h_{fd}^{A_3} + (b-a)\sin\gamma\right]$$

And:

$$\tilde{\alpha}_2 = -gA\left[\rho_f \left(\frac{m}{\Delta t} + c + \frac{k\Delta t}{2}\right)^{-1} + \frac{K_p}{\Pi}\right] \quad \textbf{(4.41)}$$

$$\tilde{\beta}_2 = A\left[A_p \left(\frac{m}{\Delta t} + c + \frac{k\Delta t}{2}\right)^{-1} + \frac{\tilde{C}_p}{\Pi \Sigma_p E}\right] \quad \textbf{(4.42)}$$

$$\tilde{q}_2 = \frac{1}{\Pi} K_p g A H^{A_3} + A\dot{u}_z^{A_3} + \frac{1}{\Pi} q_d^{A_3} + a\Delta t g A\left[\left(\Gamma - \frac{1}{\Pi}\right) h_{fd}^{A_3} + \sin\gamma\right] \quad \textbf{(4.43)}$$

$$-\left(\frac{m}{\Delta t} + c + \frac{k\Delta t}{2}\right)^{-1} A\left[\left(\frac{m}{\Delta t} - \frac{k\Delta t}{2}\right)\dot{u}_z^Q - k u_z^Q + A\rho_f g Z \sin\gamma\right] - \frac{\tilde{C}_p A}{\Pi \Sigma_p E}\sigma_z^{A_3}$$

Thus, the upstream discharge can be written with respect of the piezometric head H^P:

$$q_u^P = \tilde{\beta}\left(\frac{\tilde{\alpha}_1}{\tilde{\beta}_1} - \frac{\tilde{\alpha}_2}{\tilde{\beta}_2}\right)H^P + \tilde{\beta}\left(\frac{\tilde{q}_1}{\tilde{\beta}_1} - \frac{\tilde{q}_2}{\tilde{\beta}_2}\right) \quad \textbf{(4.44)}$$

With:

$$\tilde{\beta} = \left(\frac{1}{\Sigma_f \tilde{\beta}_1} - \frac{1}{\Pi \tilde{\beta}_1} - \frac{1}{\Pi \tilde{\beta}_2}\right)^{-1} \quad \textbf{(4.45)}$$

Consequently, the following quadratic equation can be derived thanks to Eqs. (1.41) and (1.27).

$$\left(H^P\right)^2 + \left[\tilde{Z}_L' - Z_L - H_v\right]H^P - (Z_L + H_v)\tilde{Z}_L' + \frac{\overset{*}{p_0}\alpha_0 \forall_m}{\tilde{A}_L' \rho g} = 0 \quad \textbf{(4.46)}$$

With:

$$\tilde{Z}_L' = \frac{1}{\tilde{A}_L'}\left[\forall_g^Q - 2\Delta t(1-\psi)q_u^Q - 2\Delta t\psi\tilde{\beta}\left(\frac{\tilde{q}_1}{\tilde{\beta}_1} - \frac{\tilde{q}_2}{\tilde{\beta}_2}\right)\right] \quad \textbf{(4.47)}$$

And:

$$\tilde{A}_L' = 2\Delta t\psi\tilde{\beta}\left(\frac{\tilde{\alpha}_2}{\tilde{\beta}_2} - \frac{\tilde{\alpha}_1}{\tilde{\beta}_2}\right) \quad \textbf{(4.48)}$$

The two real solutions are:

$$H_{1,2}^P = \frac{1}{2}\left(Z_L + H_v - \tilde{Z}_L'\right) \mp \left[\frac{1}{4}\left(\tilde{Z}_L' - Z_L - H_v\right)^2 + \tilde{Z}_L'\left(Z_L + H_v\right) - \frac{\overset{*}{p_0}\alpha_0 \forall_m}{\tilde{A}_L' \rho g}\right]^{1/2} \quad \textbf{(4.49)}$$

Selective Criterion

In fact, the physical conditions at a given section *i* and a time *j* lead to unique value of the head *H*. That is Eq. (4.21), (4.29) and (4.46) shall have a unique realistic solution. For instance, from mathematical view, the two solutions given by Eq. (4.24) verify the quadratic equation (4.21) but only one among them shall be retained. The retained solution of the piezometric head should be attentively selected whether calculations are performed at interior sections or at the valve. The selective criterion proposed in this study is based on the physical property of fluids to keep a minimum absolute pressure equal to zero whatever the conditions. To compare gauge pressure heads, the reference head $H_{ref} = -H_b + Z$ can be used, and three situations are involved:

i) If the two calculated solutions are greater than H_{ref}, the closest one is retained.

ii) If the two calculated solutions are lesser than H_{ref}, then the two solutions are rejected and H_{ref} is considered instead of them.

iii) If $H_1^P < H_{ref}$ and $H_2^P \geq H_{ref}$ then H_2^P is retained and H_1^P is rejected and *vice versa*.

COLUMN SEPARATION MODELLING IN VISCOELASTIC PIPES

In this section, the analysis is restrained to the use of the classical viscoelastic model to simulate cavitation. The coupled viscoelastic model referred to as the 4EVEM is not applied to simulate transient cavitation in this study. However, the classical model will take account of gas release modelled by the classical DGCM.

Use of the Classical DVCM

The application of the classical DVCM to the viscoelastic pipes uses Eqs. (3.55) and (3.56) as:

$$C_f^+ : H^P - H^A + B\left(q_u^P - q_d^A\right) + 2\Delta t \frac{C_f^2}{g}\left(\frac{\partial \varepsilon_{\varphi.r}}{\partial t}\right)^P + \Delta t C_f h_f^A = 0 \qquad \textbf{(4.50)}$$

$$C_f^- : H^P - H^B - B\left(q_d^P - q_u^B\right) + 2\Delta t \frac{C_f^2}{g}\left(\frac{\partial \varepsilon_{\varphi.r}}{\partial t}\right)^P - \Delta t C_f h_f^B = 0 \qquad \textbf{(4.51)}$$

in which the retarded strain rates are given by Eq. 3.75. It yields from the above equations. In case of vaporous cavitation, it is stated that $H = H_v - z \sin\gamma$ and the discharges can be easily derived:

$$q_u^P = q_d^A + \frac{1}{B}\left(H^A - H^P\right) - 2\Delta t C_f A\left(\frac{\partial \varepsilon_{\varphi.r}}{\partial t}\right)^P - \Delta t g A h_f^A \quad (4.52)$$

$$q_d^P = q_u^B - \frac{1}{B}\left(H^B - H^P\right) + 2\Delta t C_f A\left(\frac{\partial \varepsilon_{\varphi.r}}{\partial t}\right)^P - \Delta t g A h_f^B \quad (4.53)$$

The cavity volume is calculated according to Eq. (3.38) and the boundary conditions are the same as those of the elastic pipes.

Use of the Classical DGCM

The use of the classical DGCM to simulate cavitation in viscoelastic pipes can be obtained as the similar manner as the elastic pipes case, yet the friction term is considered more simply. Eqs. (3.55) and (3.56) allow:

$$q_d^P - q_u^P = \frac{2}{\beta B} H^P + \hat{q} \quad (4.54)$$

where the number β is defined in Eq. 3. 79 and the discharge $\hat{q} = \hat{q}_e + \hat{q}_r$ is the sum of two parts: an elastic part $\hat{q}_e$ given by:

$$\hat{q}_e = q_u^B - q_d^A + \frac{1}{B}\left(H^A - H^B\right) - gA\Delta t\left(h_f^A + h_f^B\right) \quad (4.55)$$

and a retarded part $\hat{q}_r$ obtained from Eq. 3.75 as:

$$\hat{q}_r = 2A\Delta t C_f \left\{ \begin{array}{l} -2\sum_{k=1}^{N} \frac{e^{-\Delta t/\tau_k}}{\tau_k}\left(\varepsilon_{\varphi.r}^k\right)^Q - \frac{D}{e}\rho_f g \sum_{k=1}^{N} \frac{J_k}{\tau_k} e^{-\Delta t/\tau_k} H_0 \\ + \frac{D}{e}\rho_f g \sum_{k=1}^{N}\left[\frac{J_k}{\tau_k} e^{-\Delta t/\tau_k} - \frac{J_k}{\Delta t}\left(1 - e^{-\Delta t/\tau_k}\right)\right] H^Q \end{array} \right\} \quad (4.56)$$

Incorporating Eq. (4.54) in Eq. (1.41) and then using Eq. (1.27) lead to the following quadratic equation:

$$\left(H^P\right)^2+\left[\hat{Z}-\left(Z+H_v\right)\right]H^P-\hat{Z}\left(Z+H_v\right)-\frac{p_0^*\alpha_0\forall_m}{\rho_f g\hat{A}}=0 \quad \textbf{(4.57)}$$

With:

$$\hat{A}=\frac{4\Delta t\psi}{\beta B} \quad \textbf{(4.58)}$$

$$\hat{Z}=\frac{\hat{\forall}_g}{\hat{A}} \quad \textbf{(4.59)}$$

in which:

$$\hat{\forall}_g=\forall_g^Q+2\Delta t\left[\psi\hat{q}+\left(1-\psi\right)\left(q_d^Q-q_u^Q\right)\right] \quad \textbf{(4.60)}$$

Therefore, the solution of the above equation:

$$H_{1,2}^P=\frac{1}{2}\left(Z+H_v-\hat{Z}\right)\mp\left[\frac{1}{4}\left(\hat{Z}-Z-H_v\right)^2+\hat{Z}\left(Z+H_v\right)+\frac{p_0^*\alpha_0\forall_m}{\hat{A}\rho_f g}\right]^{1/2} \quad \textbf{(4.61)}$$

As established by Wylie (1984) for the elastic pipes, the piezometric head at the point P of the viscoelastic pipe is the first solution retained in Eq. (4.61).

$$H^P=H_1^P=\frac{1}{2}\left(Z+H_v-\hat{Z}\right)-\left[\frac{1}{4}\left(\hat{Z}-Z-H_v\right)^2+\hat{Z}\left(Z+H_v\right)+\frac{p_0^*\alpha_0\forall_m}{\hat{A}\rho_f g}\right]^{1/2} \quad \textbf{(4.62)}$$

Use of the Coupled DVCM

The application of the DVCM in calculation of column separation in viscoelastic pipelines can follow the same approach proposed for the elastic pipes, but the governing equations are the integration of the compatibility equations of the 4EVEM given by Eqs. 3.97 to 3.100. It has been established that vaporous cavitation occurs if the pressure drops below the gauge vapour p_v, which corresponds to $H = H_v - z\sin\gamma$. The rearrangement of Eqs. 3.97 to 3.100 leads to the constitutive equations of the coupled VE-DVCM.

$$\frac{1}{A}q_u^P + \Sigma_f \dot{u}_z^P - \frac{\Sigma_f}{\rho_p \tilde{C}_f}\sigma_z^P = b\Delta t g\left(\Sigma_f \Gamma - 1\right)h_f^{A_1} + b\Delta t \Sigma_f g \sin\gamma +$$

$$\frac{1}{A}q_d^{A_1} - K_f g\left(H^P - H^{A_1}\right) + \Sigma_f \dot{u}_z^{A_1} - \frac{\Sigma_f}{\rho_p \tilde{C}_f}\sigma_z^{A_1} + b\Delta t \hat{C}_1 \frac{\partial\left(\varepsilon_{\varphi r}\right)^P}{\partial t} \tag{4.63}$$

$$\frac{1}{A}q_d^P + \Sigma_f \dot{u}_z^P + \frac{\Sigma_f}{\rho_p \tilde{C}_f}\sigma_z^P = b\Delta t g\left(\Sigma_f \Gamma - 1\right)h_f^{A_2} + b\Delta t \Sigma_f g \sin\gamma$$

$$+\frac{1}{A}q_d^{A_2} + K_f g\left(H^P - H^{A_2}\right) + \Sigma_f \dot{u}_z^{A_2} + \frac{\Sigma_f}{\rho_p \tilde{C}_f}\sigma_z^{A_2} - b\Delta t \hat{C}_1 \frac{\partial\left(\varepsilon_{\varphi r}\right)^P}{\partial t} \tag{4.64}$$

$$\frac{1}{A}q_u^P + \Pi \dot{u}_z^P - \frac{\tilde{C}_p}{\Sigma_p E}\sigma_z^P = a\Delta t\left(\Pi\Gamma - \Sigma_p\right)g h_f^{A_3} + a\Delta t \Pi g \sin\gamma$$

$$+\frac{1}{A}q_d^{A_3} - K_p g\left(H^P - H^{A_3}\right) + \Pi \dot{u}_z^{A_3} - \frac{\tilde{C}_p}{\Sigma_p E}\sigma_z^{A_3} + a\Delta t \hat{C}_2 \frac{\partial\left(\varepsilon_{\varphi r}\right)^P}{\partial t} \tag{4.65}$$

$$\frac{1}{A}q_d^P + \Pi \dot{u}_z^P + \frac{\tilde{C}_p}{\Sigma_p E}\sigma_z^P = a\Delta t\left(\Pi\Gamma - \Sigma_p\right)g h_f^{A_4} + a\Delta t \Pi g \sin\gamma$$

$$+\frac{1}{A}q_u^{A_4} + K_p g\left(H^P - H^{A_4}\right) + \Pi \dot{u}_z^{A_4} + \frac{\tilde{C}_p}{\Sigma_p E}\sigma_z^{A_4} - a\Delta t \hat{C}_2 \frac{\partial\left(\varepsilon_{\varphi r}\right)^P}{\partial t} \tag{4.66}$$

A constant inverse solution matrix can be used to rearrange the above equations is presented in appendix D, so that the calculation of the coupled VE-DVCM can be simply obtained.

CONCLUSION

In this chapter, numerical models of column separation in straight pipelines with FSI have been developed. The models have been applied to both elastic and viscoelastic pipelines. The MOC with WSA scheme has been basically used, since the present analysis had been based on the compatibility equations of water hammer previously described in the third chapter. The coupled DVCM has been applied to the viscoelastic pipelines. Moreover, gas release phenomenon has been taken into account by providing the coupled DGCM. This latter has been applied to the elastic pipelines, but development in case of viscoelastic pipelines seems to be more complicated. Furthermore, the valve anchoring conditions has been also considered for the coupled numerical models. For each model, the boundary condition at the valve has been considered in two different modes. The first one corresponds to the fixed valve which ignores junction coupling. The second which is more complicated corresponds to the freely moving valve. This latter supposes that the valve can move axially so that the material of the support (whether rigid or viscoelastic) matters. The implementation of the numerical models into the software package using matrix transformation is detailed in appendix D.

CHAPTER 5

Numerical Results for Elastic Pipelines

Abstract: This chapter deals with the numerical results obtained from the simulation of fluid transients in elastic pipes. The MOC with WSA scheme is used and the numerical results are validated against experimental records. The comparison shows good agreement between the two results which validate the earlier proposed models.

Keywords: Experimental result; Fluid response; Poisson coupling; Structural response; Unsteady friction.

INTRODUCTION

Water hammer is simulated by use of the 4EM and the 4EFM with a comparison between the two models. In the former, Poisson coupling, and junction coupling can be revealed, while the latter takes account of the three dynamic coupling types. In addition, the unsteady friction (UF) effect is revealed by simulating the Zielke model as well as the Vardy-Brown model in both classical and coupled models. For cavitation, two coupled column separation models, namely the coupled DVCM and the coupled DGCM are proposed and compared against experimental results from the literature.

WATER HAMMER WITHOUT CAVITATION

It is worth noting that the negative values of pressure displayed in the following figures do not simulate the real responses of the fluid, because the relative pressure does not drop below -1.013 bar. The aim of this numerical simulation is to compare the WSA scheme to the linear interpolation schemes.

The Benchmark Problem A

The Benchmark Problem A (BPA) shown by Fig. (**5.1**) is used as numerical test and the exact solution of Tijsseling [2] is used to compare and evaluate the WSA scheme. Noting that the Delft Hydraulic Problems A to F have been defined and used to test numerical methods and FSI software. The BPA concerns a reservoir-pipe-valve system, where the straight steel-pipe is defined by its length $L = 20$ m, inner radius $R = 398.5$ mm, thickness $e = 8$ mm, Young's modulus $E = 210$ GPa,

Abdelaziz Ghodhbani, Ezzeddine Haj Taïeb, Mohsen Akrout & Sami Elaoud

mass density $\rho_p = 8900$ kg.m^{-3} and Poisson coefficient $\nu = 0.30$. The water in the pipe has a bulk modulus of elasticity K = 2.1 GPa, mass density $\rho_f = 1000$ kg.m^{-3} and an initial velocity $V_0 = 1$ m.s^{-1}. The instantaneously closing valve may be fixed (no junction coupling) or free (with junction coupling). Tijsseling proposed and validated [80] the exact solution for the BPA against the MOC solution obtained with the WSA scheme proposed [81]. The calculation shown that the two solutions are almost identical.

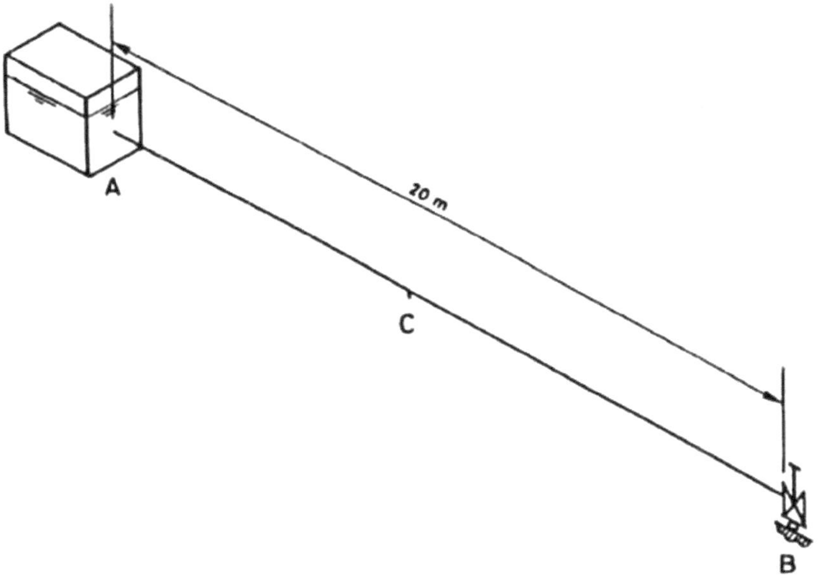

Fig. (5.1). Reservoir-pipe-valve system in Delft Hydraulics Benchmark Problem A [80].

Effect of Poisson Coupling

This subsection concerns the analysis of the Poisson coupling assumed to act without junction coupling: the valve is rigidly fixed to the support. The WSA scheme of the MOC is used to calculate the 4EM. The characteristic directions are $\tilde{C}_f = 1024{,}55$ m.s^{-1} and $\tilde{C}_p = 5280{,}5$ m.s^{-1} so that their ratio is $\tilde{C}_p / \tilde{C}_f = 5.1539$. Therefore, the integers $a = 13$ and $b = 67$ are chosen for the adjustment. Noting that the computations are obtained without any correction on the mass densities, because the result is practically not influenced with these assumptions. Also, there is no discretization of the space domain; only one reach can be used, *i.e.* $N = 1$ and $\Delta t = 0.98$ ms. To calculate the pressure at the middle of the pipe, a minimum number of reaches $N = 2$ will be sufficient. The effect of Poisson coupling on the pressure evolution at the valve is shown by Figs. (**5.2** and **5.3**). These figures also display the preference of the WSA scheme to both SLI and TLI schemes. The effect of the precursor wave on the pressure wave can be clearly observed. The precursor waves are pressure waves induced by the stress waves and propagate at the stress wave-speed C_p.

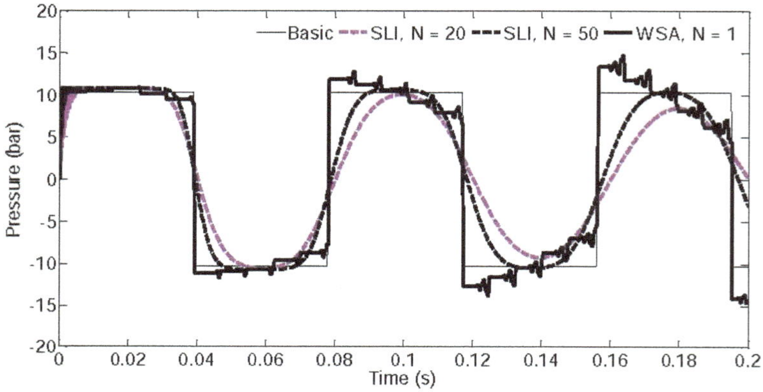

Fig. (5.2). Pressure at the fixed valve and preference of the WSA scheme to the SLI scheme in the BPA.

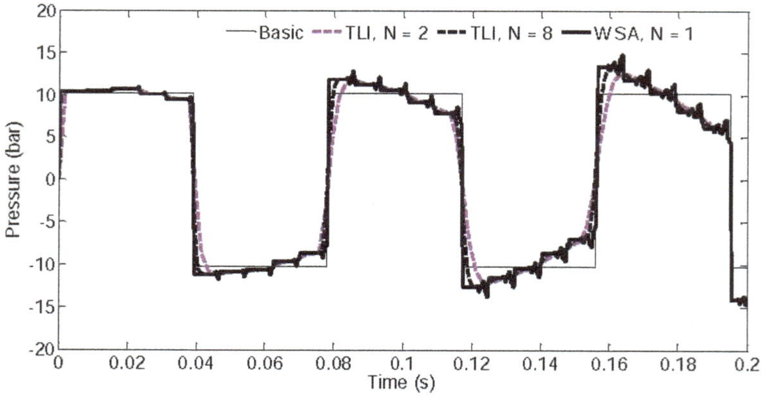

Fig. (5.3). Pressure at the fixed valve and preference of the WSA scheme to the TLI scheme in the BPA.

Fig. (**5.4**) illustrates the precursor wave where the WSA is compared against the TLI solutions. The magnitude of the precursor wave is the same for the two schemes and almost equal to 0.1 bar. The fact that the characteristic direction ratio is close to $67/13$ means that 67 stress waves fit in 13 pressure waves. This ratio can be directly measured on the graphic by calculating the ratio of the front wave positions. The front waves of both the fluid and the structure exist, respectively at 17 m and 3.3 m up to the reservoir, so that their ratio is 5.15, which is close to 5.1539. The reflected precursor waves as well as the axial stresses (Fig. **5.5**) affect the pressure shape at the valve as shown in Figs. (**5.2** and **5.3**). The precursor wave accumulation in time causes the beat phenomenon [14]. Fig. (**5.6**) displays the beat phenomenon in the no-junction coupling case and by use of the WSA scheme.

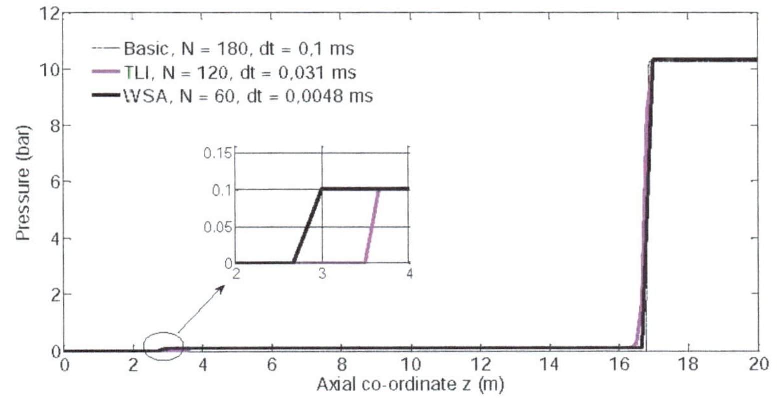

Fig. (5.4). Precursor wave without junction coupling in the BPA, shown at time t = L/C_f = 3.25 ms

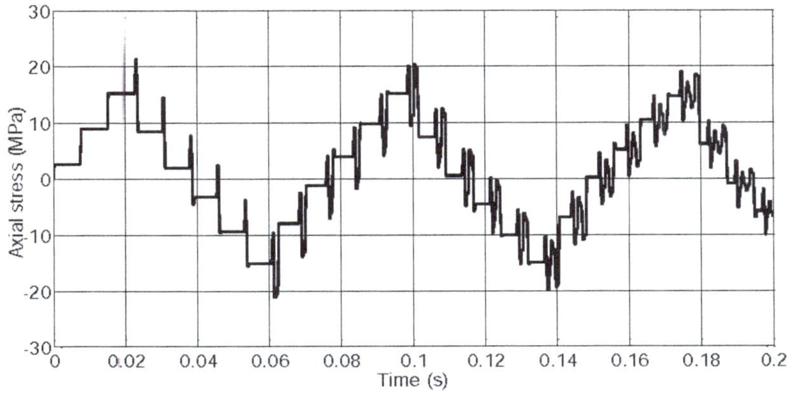

Fig. (5.5). Axial stress at the fixed valve in the BPA, calculated with the WSA scheme.

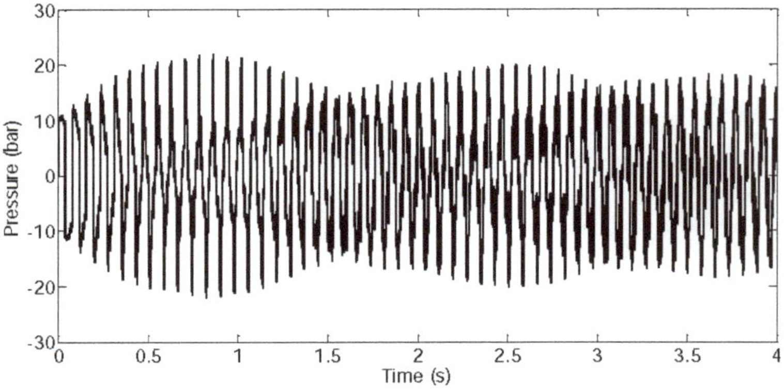

Fig. (5.6). Simulation of the beat phenomenon at the fixed valve in the BPA, using the WSA scheme.

Effect of Junction Coupling

For the reservoir-pipe-valve system (BPA case), junction coupling corresponds to the unrestrained valve. The WSA is again preferred to both SLI and TLI schemes (Figs. **5.7** and **5.8**).

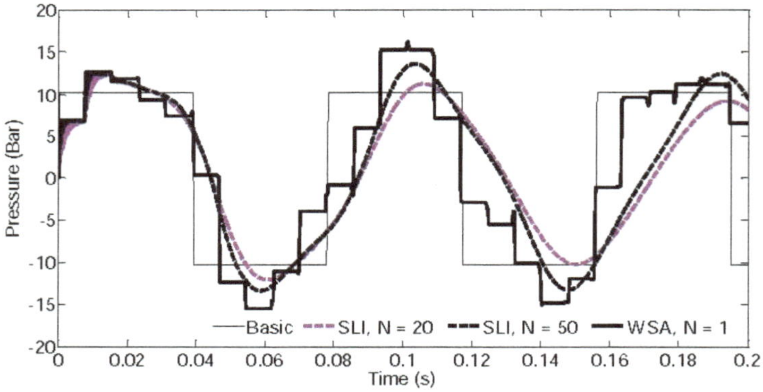

Fig. (5.7). Pressure at the free valve in the BPA and preference of the WSA scheme to the SLI scheme.

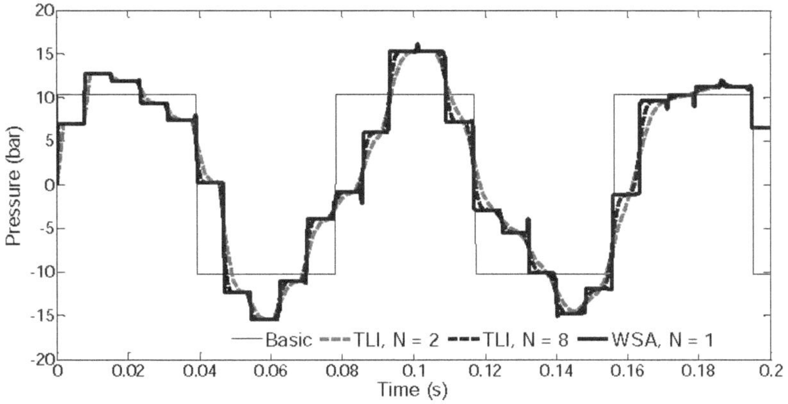

Fig. (5.8). Pressure at the free valve in the BPA and preference of the WSA scheme to the TLI scheme.

The period of the beat phenomenon denoted can be measured on the graphic. It is almost equal to 1.6 s. Tijsseling proposed a formula to calculate with respect to the water hammer period T and the characteristic direction ratio r as [81]. For T = 0.078 s, the calculated period is of about 1.6 s. Nevertheless, the shapes of the pressure plots are different from those of the fixed valve case. This is due to the higher magnitude of the precursor wave almost equal to 0.55 bar (Fig. **5.9**). The difference between the two configuration is greater which demonstrates the strong effect of junction coupling.

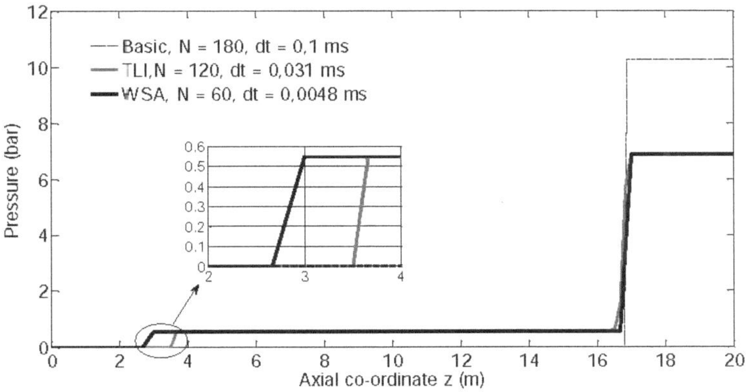

Fig. (5.9). Precursor wave with junction coupling in the BPA, shown at time t = 3.25 ms.

The axial stress diagram shown by Fig. (**5.10**) has the same shape as the pressure diagram, since the axial stress is linearly dependent to the pressure unlike the no-junction coupling case. The beat phenomenon is also influenced by this higher precursor wave as shown in Fig. (**5.11**).

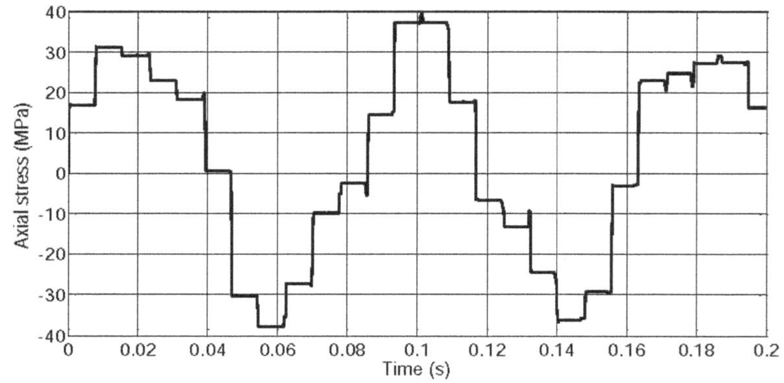

Fig. (5.10). Axial stress at the free valve in the BPA.

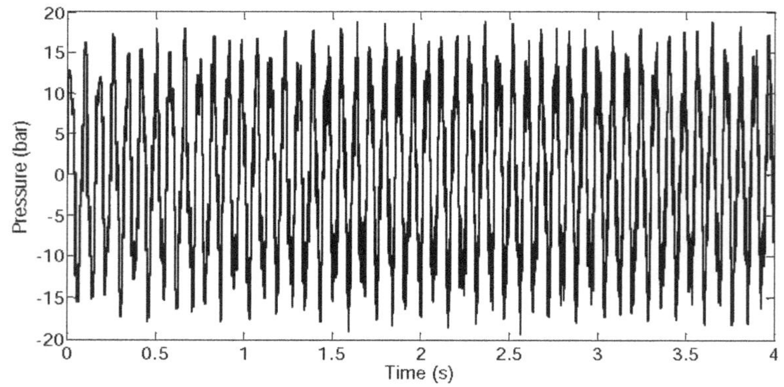

Fig. (5.11). Simulation of the beat phenomenon at the free valve in the BPA using the WSA scheme.

Effect of Friction Coupling

The 4EFM is solved by the MOC using regular mesh with space step Δz and time step Δt. Since the steady-state flow in the experiment of Haj Taïeb is laminar (Re = 1961), the Zielke model can be used. To validate the proposed model, the pressure at the free moving valve is calculated with three numerical schemes and compared against measurement.

Figs. (**5.12** and **5.13**) show insufficiency of the previous models (Zielke model and 4EM) to predict water hammer phenomenon in case of junction coupling. Fig. (**5.12**) shows correct simulation of damping using the Zielke model, whereas shape and timing are incorrectly simulated. The timing disagreement is due to the difference between the pressure wave celerity based on experimental data almost equal to 1150 $m.s^{-1}$ and the calculated celerity (given by the Korteweg's formulae) which is close to 1243.6 $m.s^{-1}$. According to the experimental specifications, the pressure of the reservoir is equal to 2.6 bar and the Joukowsky pressure change is $\Delta p = 1.02$ bar. However, the calculated pressure rise reaches 4.5 bar. The exceedance of 0.88 bar may be due to the bounding shape of the pipe. It is worth noting that the preciction of the first pressure rise induced by water hammer phenomenon is of big importance in pipeline integrity because the determination of safety components depends on it [59]. The Zielke model gives a first pressure rise equal to the Joukowsky pressure rise and regular damping at the other peaks. Fig. (**5.13**) shows the simulation of the pressure history at the valve using the 4EM where a maximized number of reaches is used for each scheme. It- can be observed that the computations obtained by the 4EM failed to give accurate prediction of the first pressure rise. Although the calculated value exceeds the Joukowsky pressure change, the difference between measurement is equal to 0.4 bar in cases of SLI scheme and WSA scheme and 0.3 bar for the TLI scheme. Noting that the quasi-steady approach gives pressure peaks which exceed the maximum measured pressure, but they cannot be considered as good results since they are unrealistic data.

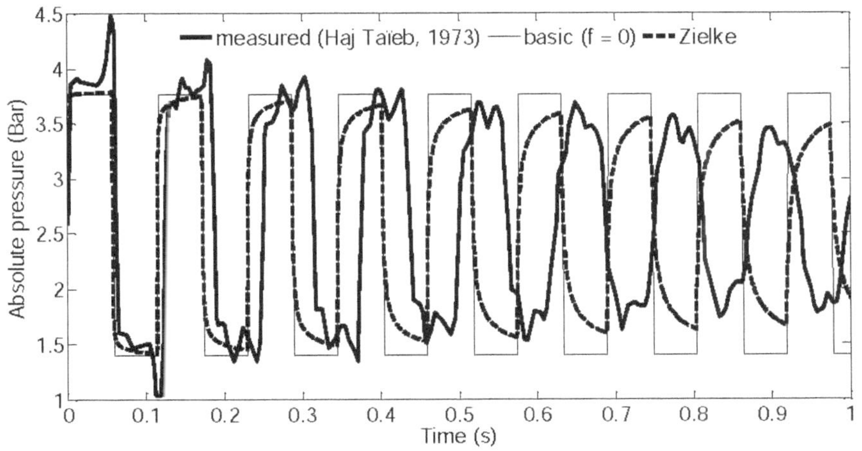

Fig. (5.12). Pressure history at the valve showing insufficiency of the Zielke model.

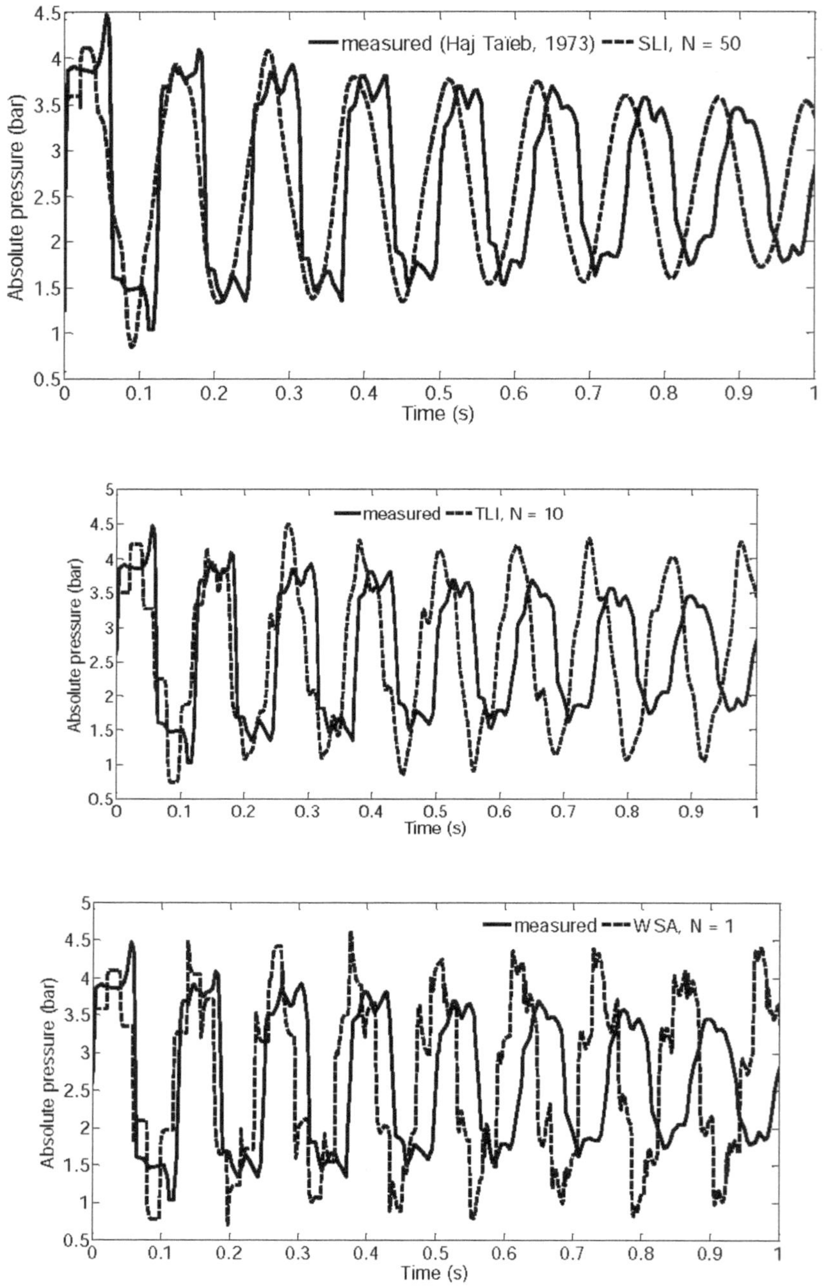

Fig. (5.13). Pressure history at the valve showing insufficiency of the 4EM.

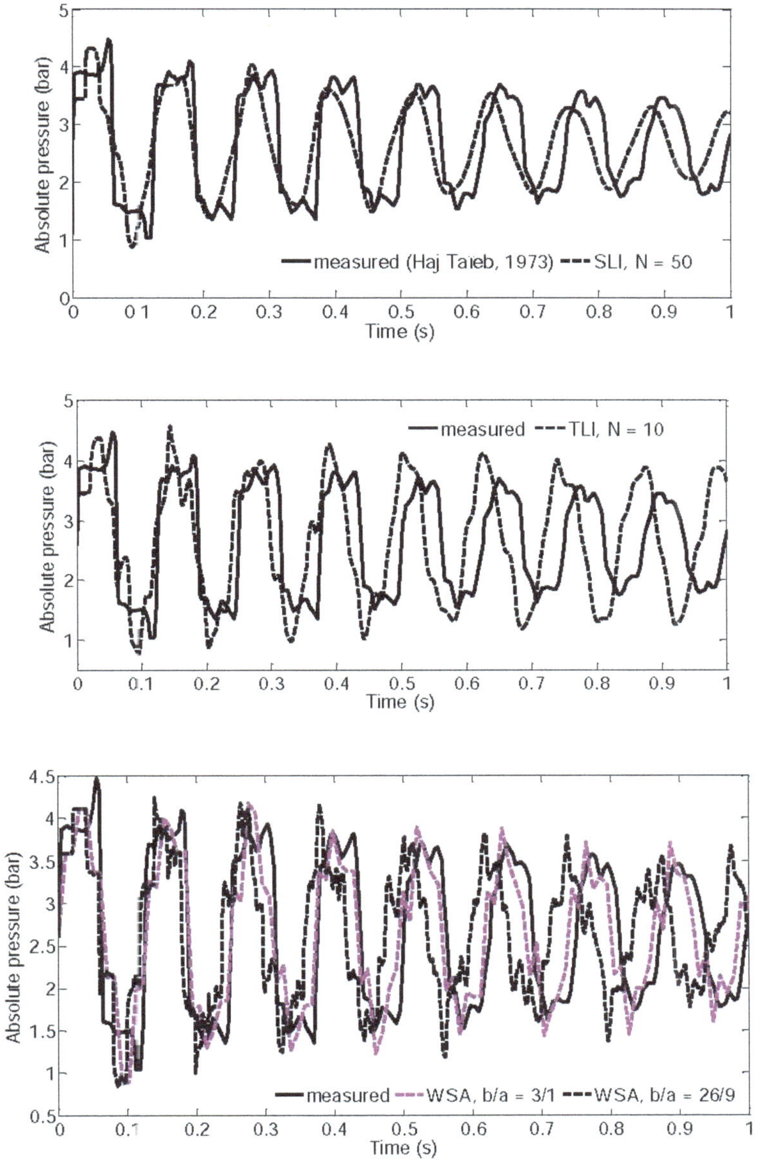

Fig. (5.14). Pressure history at the valve showing efficiency of the 4EFM.

Fig. **(5.14)** proves the efficiency of the 4EFM to predict water hammer. Stability and convergence are good for the three numerical schemes. Furthermore, numerical errors are decreased, and the curve shape is improved. Although the timing error is still existent, damping is nicely corrected with SLI scheme and WSA scheme. The first pressure rise is more improved with the SLI scheme, but its main disadvantage is the relatively long computation time compared to the other schemes. The TLI scheme still shows unrealistic pressure peaks even with small time step ($N = 50$). The WSA gives acceptable results, and it needs a short computation time. The calculations give a characteristic direction ratio close to 2.8935. Several integers *a* and *b* (with $a < b$) can give rational numbers around this value, such as $29/9$ $23/8$, $3/1$, *etc.* It is noted that the accuracy of the WSA scheme is strongly dependent on the integers *a* and *b* used in calculation. The smaller the integers *a* and *b*, the smaller the damping error. Moreover, numerical oscillation can also be reduced by the uses of refined time steps. Stability and convergence of this scheme is excellent and accurate solution can be kept with small number of elements.

ACTIVE COLUMN SEPARATION FLOW REGIME

This section deals with the prediction of column separation accompanied by high pressure pulse due to cavity collapse. This cavitation corresponds to the low velocity case resulting in a pressure rise larger than the Joukowsky valve closure pressure. This phenomenon is called "Active column separation flow regime" [82]. The coupled DVCM and the DGCM are validated against experimental results. The experiment of Bergant and Simpson described in the first chapter is used to validate the proposed models. The experiment exhibits numerous case studies of column separation involving the pipe slope, the reservoir head, the valve location, and the initial velocity [66]. In the present work, the computations are performed for a rapid-closure downstream valve and an upward sloping pipe.

Computational Parameters

The head of the upstream tank 2 is maintained at $HT_2 = 22$ m. The measured pressure wave-speed is $C_f = 1319$ m.s^{-1} [82]. To assess numerical results performed by the classical DVCM, the calculated wave-speed $C_f = 1323$ m.s^{-1} is used instead of the measured one. The low-velocity case corresponds to $V_0 = 0.3$ m.s^{-1} and the corresponding Reynolds number is $\mathrm{Re} = 5970$ [25]. The Vardy-Brown's model can be used for UF modelling. Since the absolute roughness of copper is in the range from 0.001 mm to 0.002 mm, the maximum relative roughness ε/D of the pipe wall in the experiment of Bergant and Simpson [66] is equal to 9×10^{-5}. This value corresponds to a limit in the rough pipe condition

(Appendix E). Thus, both rough pipe and smooth pipe assumptions can be adopted in the Vardy-Brown model. Since the kinematic viscosity at the pipe wall is not available in the literature, the laminar viscosity is used. To end with, in all algorithms, the cavity volume is calculated by taking a weighting factor $\psi = 1$.

Contribution of the Classical Models

Simulation of the Classical DVCM

The numerical results provided by the classical DVCM are assessed. Figs. (**5.15** and **5.17**) show the insufficiency of this model to predict the piezometric head at the valve. The computations are down in rectangular (RG) and staggered (SG) where pressure wave-speed is assumed to be constant. Fig. (**5.15**) shows that the use of SF yields inaccurate solution. In addition, the RG causes numerical oscillations over the pressure diagram, which are avoided with the SG. The timing delay shown for the two computaional grids is slightly decreasing with mesh refinement, but it still remains for long transient duration. The two computational grids converge to the same solution for the first and the second pressure rises, but a cavity opening advance is still existent. Regarding the pressure amplitude, the two grids involve errors exhibited for the pressure peak of almost 102 m against 95 m in the experimental runs.

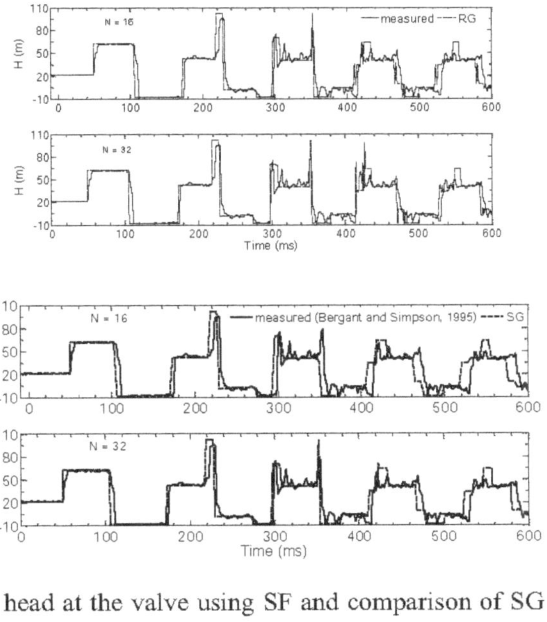

Fig. (5.15). Piezometric head at the valve using SF and comparison of SG and RG in case of the classical DVCM.

The effect of UF is shown in Figs. (**5.16** and **5.17**) where SG and RG are compared. The smooth pipe solution displayed in Fig. (**5.16**) has less numerical oscillations, but it involves more timing delay. In contrast, the rough pipe solution (Fig. **5.17**) is in good agreement with measurement regarding frequency, but it exhibits more unrealistic oscillations which increases-with the mesh refinement. Whether the pipe is smooth or rough, the SG solution seems more correct than the RG solution. The duration of the short pressure pulse simulated is about 5, which is much shorter than the experimental one almost equal to 20 ms. Despite that, the classical DVCM solved in SG with UF can reduce the maximum pressure spike to 96 m.

The piezometric head at the middle of the pipe is predicted and shown in Fig. (**5.18**), where SG is preferred for the classical DVCM. Both simulation and experimental result show cavity formation. UF is evaluated for smooth pipe and rough pipe assumptions, and the comparison is the same as in Figs. (**5.16** and **5.17**). For the two cases, the duration of the first cavity, almost equal to 25 ms, is lesser than that at the valve. Furthermore, the cavity collapse involves a pressure peak of almost 70 m, which is larger than the Joukovsky pressure rise almost equal to 62 m. For the smooth pipe solution, the calculated and experimental result are in good agreement until occurrence of the second pressure spike at $t = 260$ ms . From that time on, low-amplitude pressure pulses cannot be simulated with the classical DVCM. In addition, an increasing timing difference is also shown for the third, the fourth and the fifth pressure rises. Nonetheless, these discrepancies do not exist in the rough pipe solution where timing agrees with measurement, but the result shows unrealistic pressure spikes.

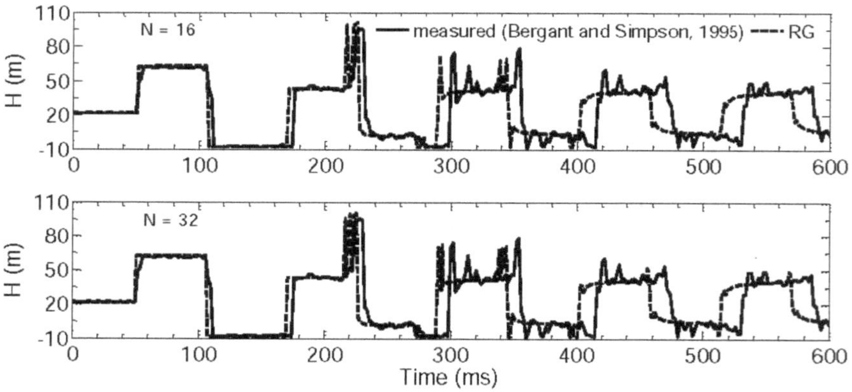

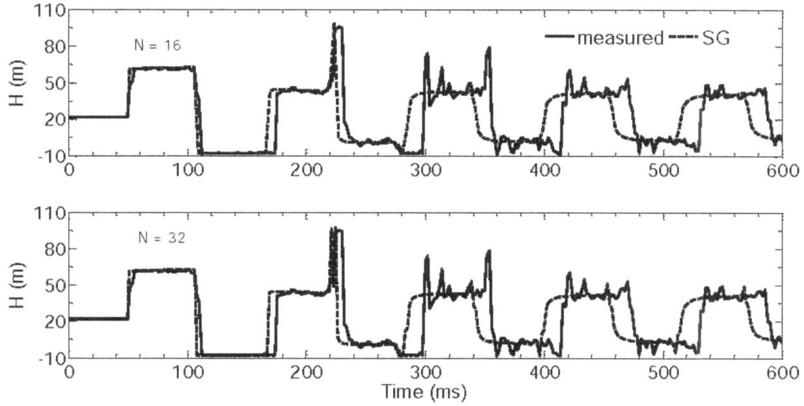

Fig. (5.16). Piezometric head at the valve using UF in smooth pipe and comparison of SG and RG in case of the classical DVCM.

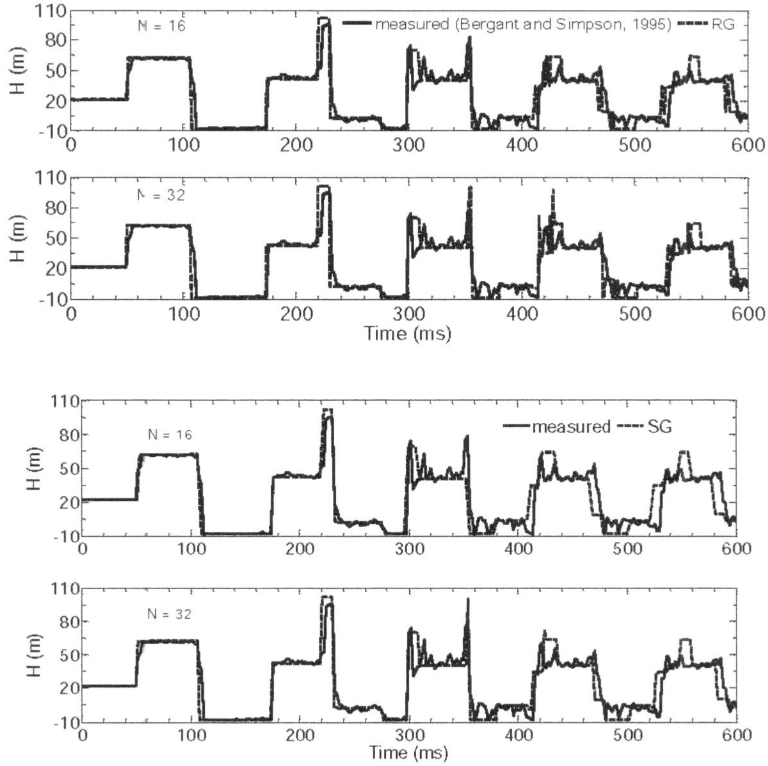

Fig. (5.17). Piezometric head at the valve using UF in rough pipe and comparison of SG and RG in case of the classical DVCM.

According to the previous simulation shown in Figs. (**5.15** and **5.17**) 5, one can conclude that friction coupling has important impact on the pressure evolution. UF is preferred than SF despite the longer computational time that it involves. However, it is still difficult to accurately model this UF effect. In fact, some industrial pipes are neither smooth nor rough, like the copper pipe used in the experiment of Bergant and Simpson. Consequently, the modelling of their roughness is more difficult, and the numerical result is less accurate.

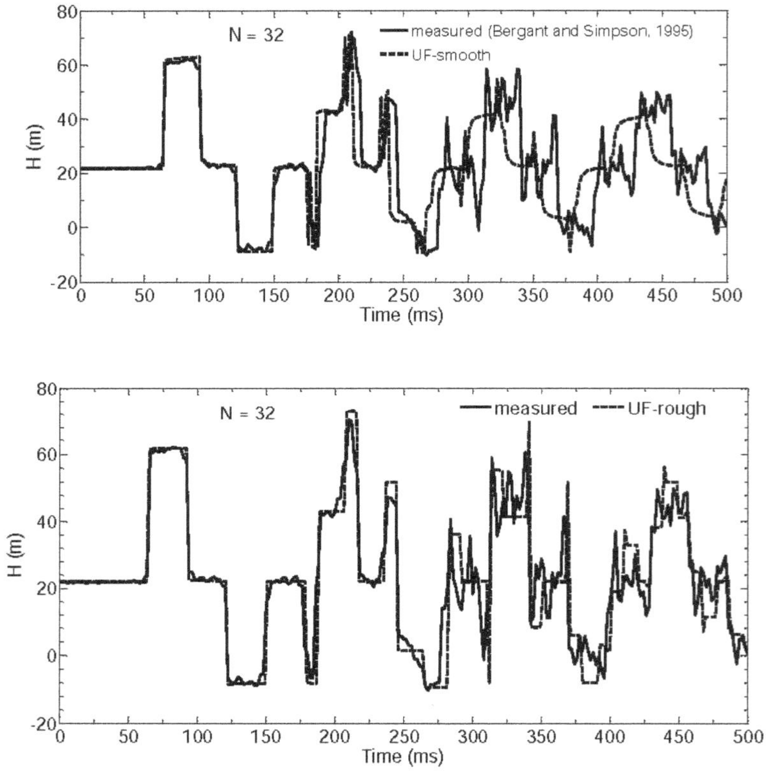

Fig. (5.18). Piezometric head at the middle of the pipe calculated with UF in SG in case of the classical DVCM.

Simulation of the Classical DGCM

The results provided by the classical DGCM are assessed. Figs. (**5.19** and **5.20**) show the efficiency of the classical DGCM in prediction of column separation in low-velocity case for which the weighting factor ψ is equal to 0.5. (Figs. **5.21** and **5.22**) exhibit the effect of ψ on the numerical result for UF in smooth pipe and for a relatively high number of reaches ($N = 32$). It is observed that the two curves are identical until the time $t = 290$ ms . Beyond this time, the calculated result obtained for $\psi = 0.5$ agrees better with the experimental result regarding- timing. However, unrealistic pressure spikes cannot be reduced. Theses spikes are, in contrast, reduced for , but a slight timing delay is involved. The effect of the pipe roughness previously discussed for the classical DVCM is still similar for the classical DGCM. Nevertheless, the weighting factor involves more numerical instability. Basically, this factor should be equal to 0.5, but because of the numerical errors, it is recommended to use 1 instead. Regardless the weighting factor, it is observed that when gas release effect is considered, the numerical result is improved. The classical DGCM is then preferred to the classical DVCM, especially in case of fluid transients with high-density dissolved air.

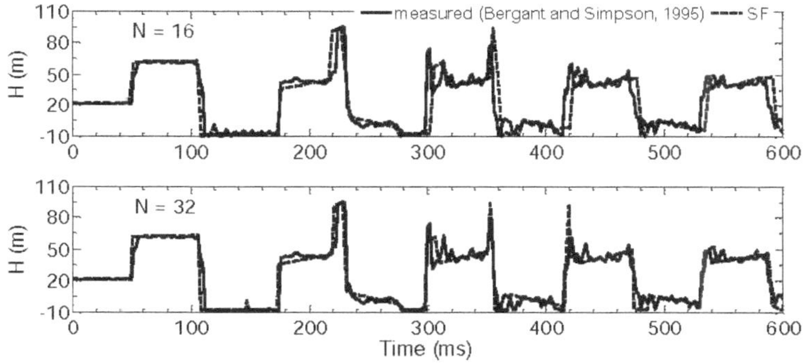

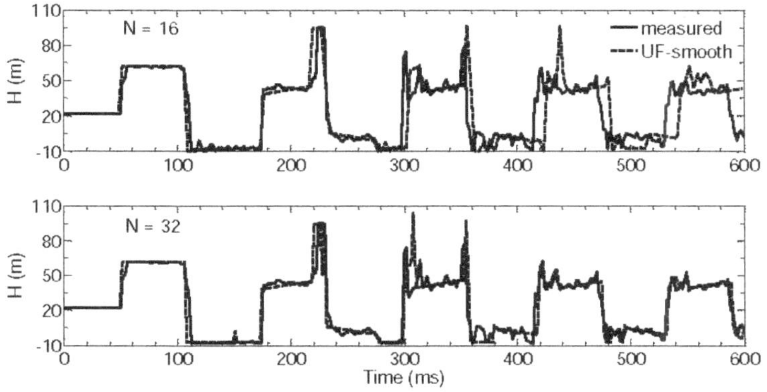

(Fig. 5.19) contd.....

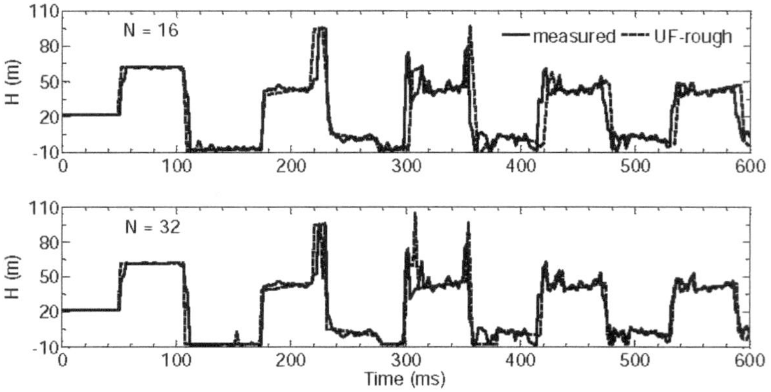

Fig. (5.19). Piezometric head at the valve for $\psi = 0.5$ obtained by the classical DGCM.

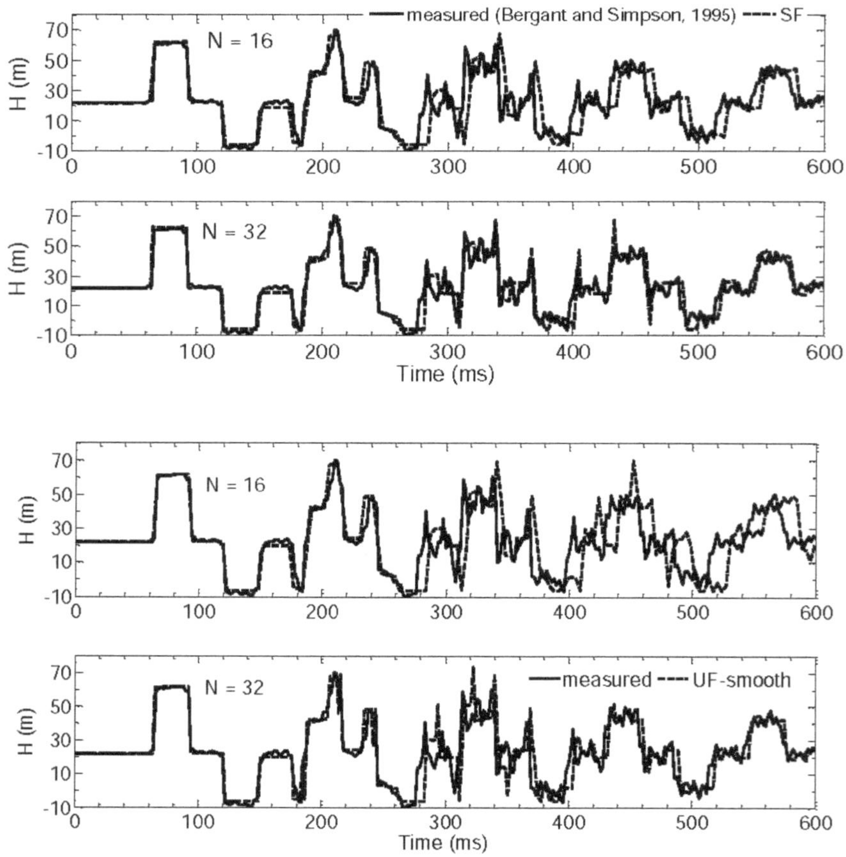

(Fig. 5.20) contd.....

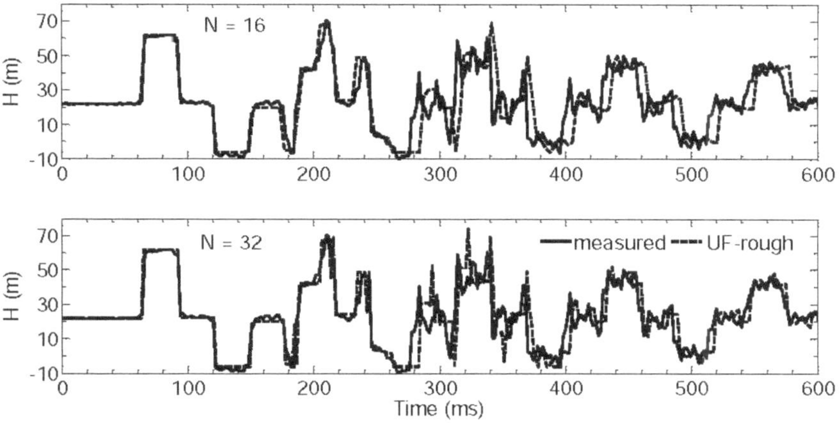

Fig. (5.20). Piezometric head at the midpoint for $\psi = 0.5$ obtained by the classical DGCM.

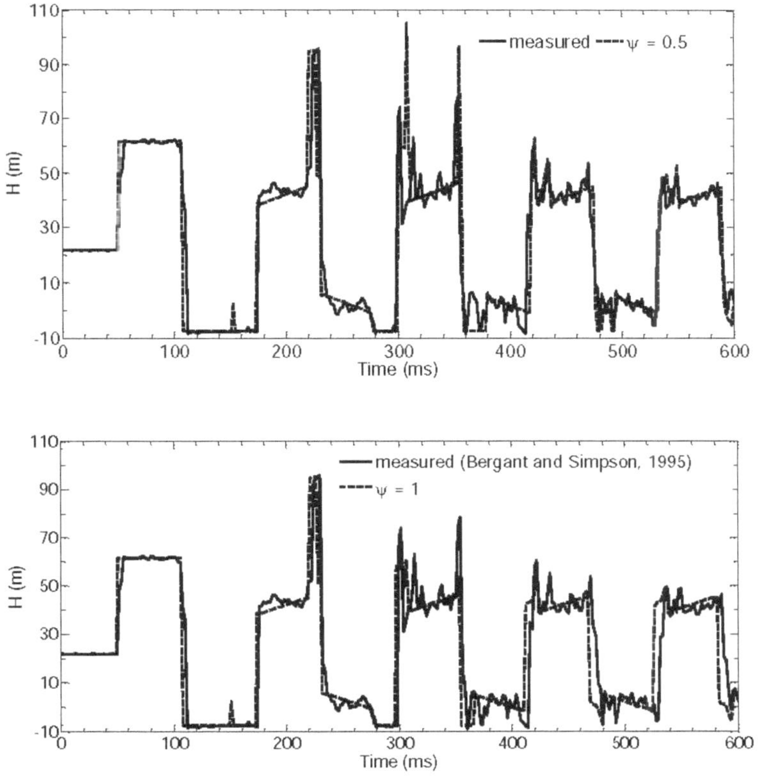

Fig. (5.21). Piezometric head at the valve showing the effect of the weighting factor ψ on the calculated piezometric head for UF in smooth pipe in case of the classical DGCM.

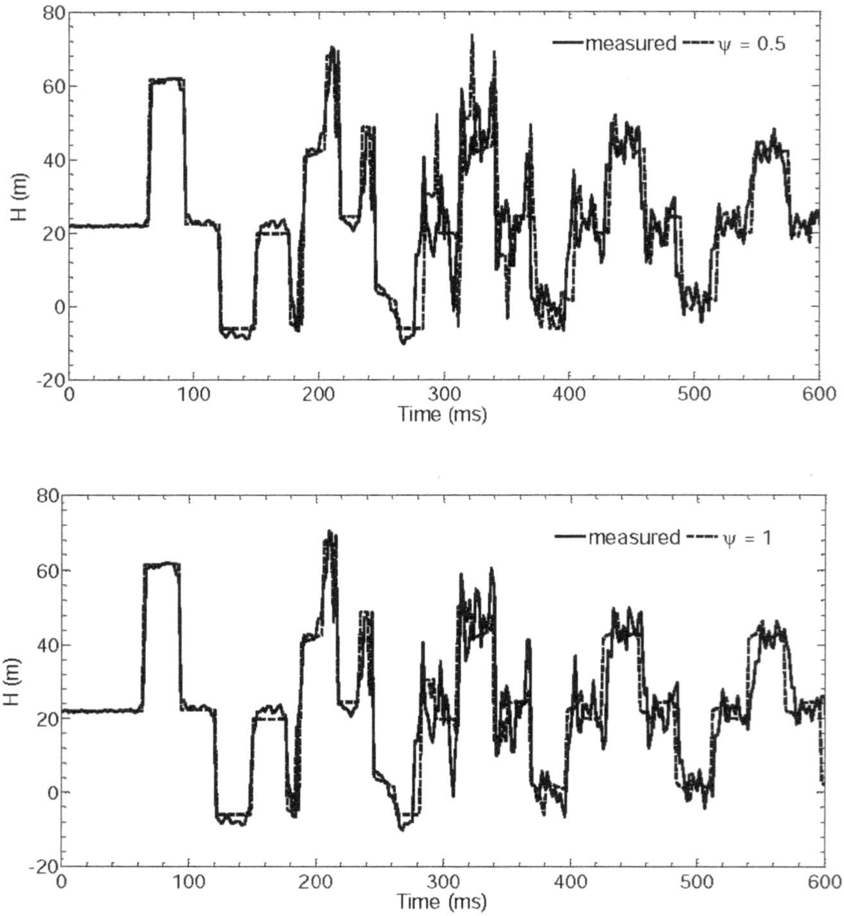

Fig. (5.22). Piezometric head at the midpoint showing the effect of the weighting factor ψ on the calculated piezometric head for UF in smooth pipe in case of the classical DGCM.

Simulation of the Coupled DVCM

The experiment of Bergant and Simpson (1995) presented in the first chapter is used again for simulation of the coupled DVCM. Unfortunately, there is no information about the axial vibration of the pipe. In a first purpose, junction coupling is not considered; only Poisson coupling and unsteady friction are assessed. Then, the freely moving valve assumption will be treated and the effect of junction coupling on the solution will be studied.

Fluid Responses with Poisson Coupling

If the constant wave-speed (CWS) method is considered, the characteristic directions are, respectively, 3721.76 m.s^{-1} for the pipe and 1331.19 m.s^{-1} for the liquid, so that the ratio r is equal to 2.7957. (Figs. **5.23** and **5.26**) investigate solution of the coupled DVCM based on the CWS method with constant number of reaches $N = 4$. Different rational numbers close to the latter ratio can be used, such as $14/5$;$17/6$; $20/7$ and $23/8$.

Fig. (**5.23**) shows insufficiency of the RG to predict column separation at the valve with the coupled DVCM, even when UF term is used. The high frequency oscillations generated are due to the unrealistic cavities with large volume occurring at the valve.

Fig. (**5.24**) shows the efficiency of the SG in prediction of column separation at the fixed valve with SF. Four different rational numbers are used and compared with each other and against the experimental result. Although unrealistic oscillations are omitted, the SG involves timing delay which decreases with the increase of the rational number. This can be attributed to the nature of the WSA method used for integration of the compatibility equations of the MOC. The mesh grid used in the present calculation is a stress-wave grid. The numerical frequency of the simulated pressure wave is strongly dependent on the pressure wave-speed C_f assumed to be constant in time. The time step Δt given by Eq. (3.28) can be introduced into the computation code in two similar forms: (i) $\Delta t = \Delta z / (b\tilde{C}_f)$ or (ii) $\Delta t = \Delta z / (a\tilde{C}_p)$. For instance, if expression (i) is used, and one tries to test another rational number n slightly greater than the previous one (such as $20/7$ instead of $14/5$), then Δt decreases. Eq. (3.28) implies $\Delta z / (b\Delta t) = \tilde{C}_p / n$, with $n = b/a$, then, the term $\Delta z / (b\Delta t)$ that should be maintained equal to $\tilde{C}_f$ is slightly decreased and causes an increase in the period as shown in Figs. (**5.24** and **5.26**). Such expression is used in the present study. Similar explanation can be deduced if expression (ii) is used. In this case, Eq. (3.28) implies $\Delta z / (a\Delta t) = n\tilde{C}_f$, which means increase in the term $\Delta z / (a\Delta t)$ that should be maintained equal to the characteristic direction $\tilde{C}_p$

.

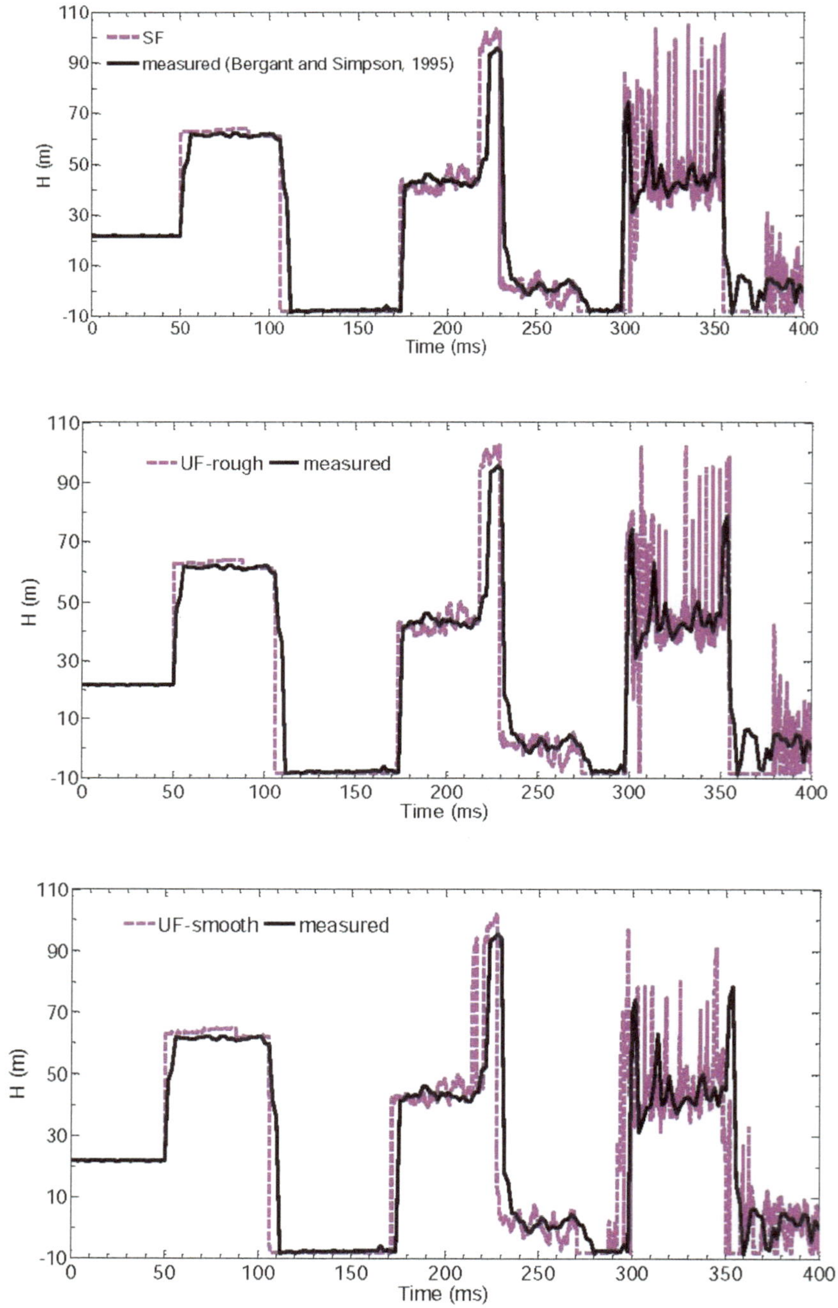

Fig. (5.23). Piezometric head at the valve calculated in RG obtained by the coupled DVCM.

N = 8

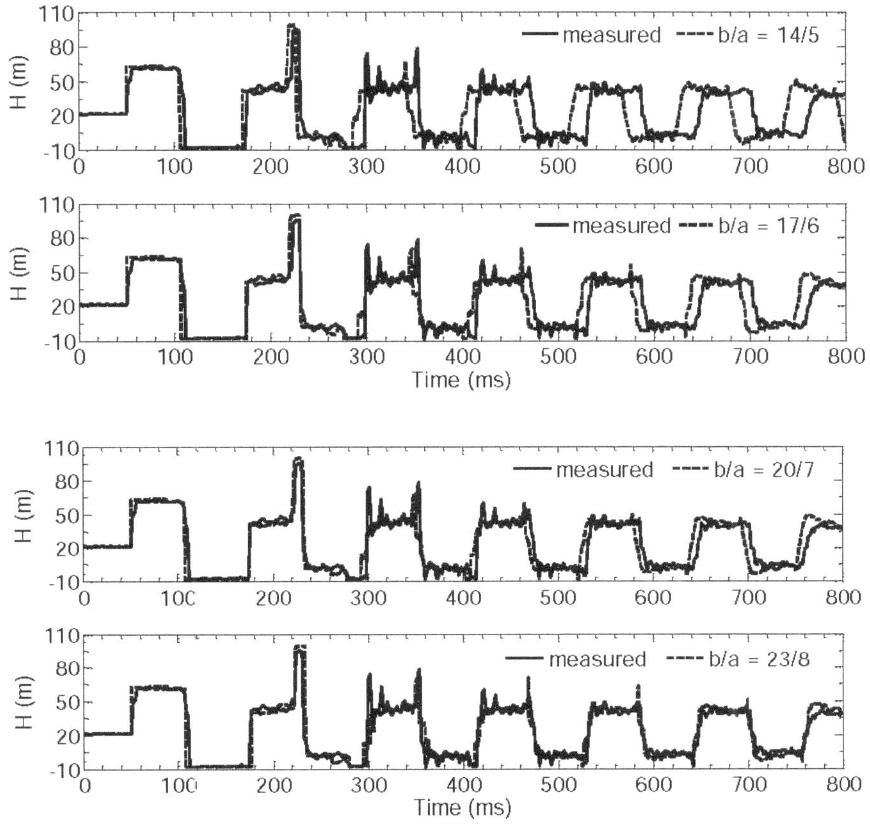

Fig. (5.24). Piezometric head at the valve using SF obtained by the coupled DVCM.

The calibration of the rational number is obtained manually. However, it is possible to make computer programs allowing automatic calibration. For example, the selection of the adequate integers may depend on how the rational number is close to the characteristic-directions ratio. Another alternative is to compare the magnitudes of numerical errors, as shown in Fig. **(5.24)**; 23/8 is more suitable regarding timing, but 20/7 involves less numerical errors.

The main advantage of the coupled DVCM is the lesser computational time; a small number of reaches $N = 4$ has been used in the previous calculation and has provided accurate results. The convergence of the solution is improved compared to the classical models. Moreover, damping effect is accurately predicted by use of the SG even when SF is used. UF damping with smooth pipe is slightly more accurate than rough pipe damping. (Figs. **5.25** and **5.26**).

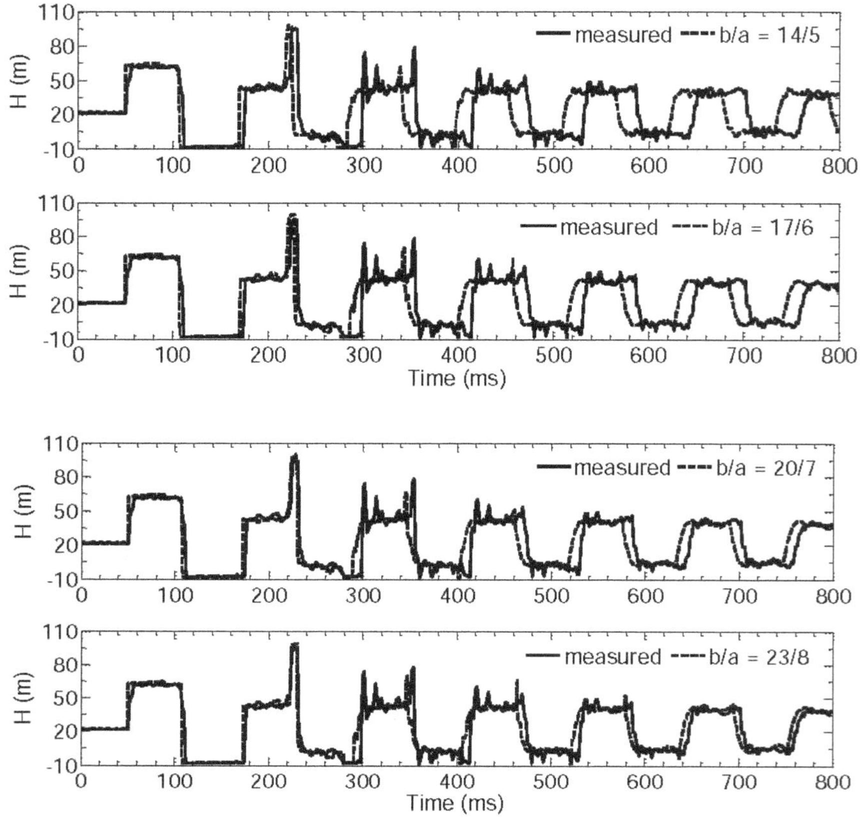

Fig. (5.25). Piezometric head at the valve using UF in smooth pipe obtained by the coupled DVCM.

To improve the result, attenuation of magnitudes can be adjusted by use of the variable wave-speed method (VWS). The damping discrepancies are due to the reduction of the pressure wave-speed in saturated liquid overlying the adjacent reaches. The pressure dependence wave-speed can be explained by the fact that the use of greater rational numbers in the CWS method, such as 20/7 and 23/8 leads to more agreement with experiment in regards of timing (Figs. **5.24** and **5.26**). If an isentropic process is considered in the cavitation region, then a narrow margin of vapour pressure can be shown using the p-h diagram of water (enthalpic diagram). The mass fraction of vapour is indicated by curves in the saturated zone. The minimum vapour pressure corresponding to the maximum cavity size can be estimated with the p-h diagram of water. For more details, thermodynamic tables and diagrams allowing accurate calculations of several water properties are available in [83].

N = 8

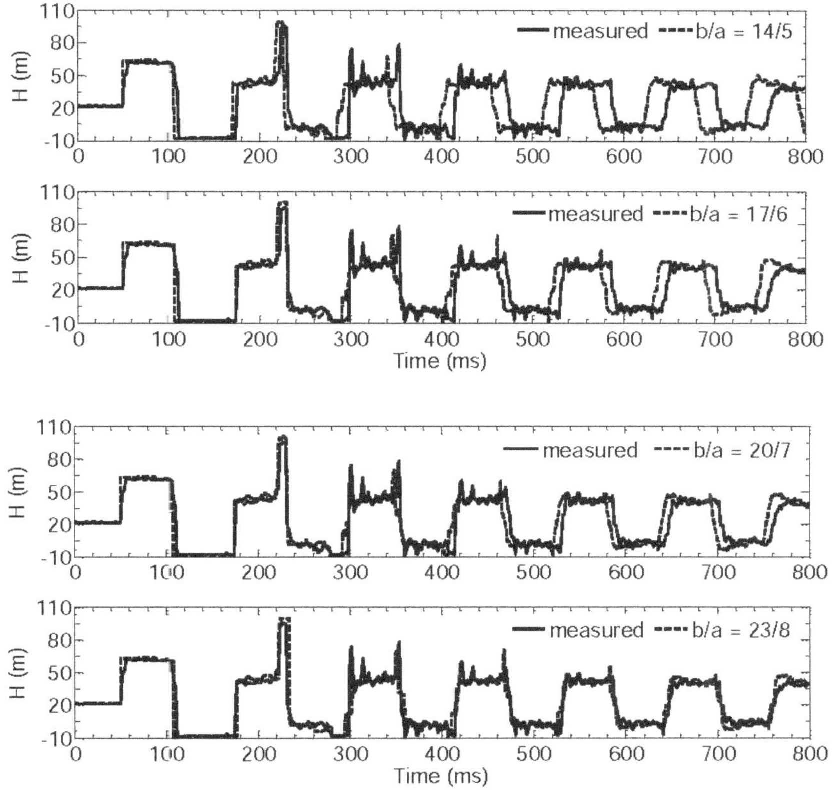

Fig. (5.26). Piezometric head at the valve using UF in rough pipe obtained by the coupled DVCM.

In this study, the starting point is the saturated state of $T = 20\ °C$ and $p = 0.02339$ bar (Table **5.1**.) The growing of the isentropic cavity leads to a small mass fraction almost equal to 2.5 %, which corresponds to a minimum vapour pressure close to 1 kPa. The saturated temperature is then deduced and subsequently, the specific volume v and the isothermal compressibility κ_T are obtained with thermodynamic tables. The density ρ_f and the bulk modulus of elasticity K of the saturated liquid are, respectively, the inverses of v and κ_T. Consequently, the pressure wave-speed is calculated according to Eq. (4.18).

Table 5.1. Physical properties of water in the vicinity of cavity.

-	Cavity Volume	
Property	**Cavity Opening**	**Maximum Volume**
p_v (bar)	0.02339	0.01
T (°C)	20	6.96963
v ($m^3.kg^{-1}$)	0.00100184	0.00100014
ρ_f($kg.m^{-3}$)	998.16	999.86
κ_T ($10^{-6}.kPa^{-1}$)	0.45836	0.48592
K (GPa)	2.18169	2.05795
C_f ($m.s^{-1}$)	1349.26	1313.49

The pressure wave-speed change shown in Table **5.1** involves changes in the characteristic directions to $3718.90\ m.s^{-1}$ and $1296.91\ m.s^{-1}$, respectively for the pipe and the fluid, so that $r = 2.8675$. Thereby, the rational number $20/7$ is used to adjust the wave-speeds and to allow calculations during the cavity forming until its collapse. The timing differences involved in the previous calculations by the classical DVCM and the coupled DVCM with the CWS method are avoided by use of the VWS method. The numerical solution and the experimental results are in precise agreement until the fifth pressure rise even with SF (Fig. **5.27**). UF in smooth pipe gives more accuracy regarding damping. In addition, the pressure peak is in excellent agreement with the experimental one unlike the SF solution. In this latter, the pressure peak exceeds the experimental measurement and reaches 102 m. Figs. (**5.28** and **5.29**) show the piezometric heads at the valve and at the middle of the pipe. The preference of the VWS method to the CWS method in case of SG and UF in smooth pipe is also observed. The first timing of both cavity opening and cavity collapse are almost the same for the two methods, and they are in good agreement with the measurement. Nevertheless, the first pressure rise prediction at the valve is excellent when the VWS method is applied to the coupled DVCM with a small number of reaches ($N = 4$), while the CWS still involve errors of almost 5 m. Moreover, the duration of the second vapour cavity formed at the valve is accurately simulated with the coupled DVCM using the VWS method. The accuracy of the following pressure spike is also improved, whereas the CWS method involves unrealistic oscillations (the errors are of almost 5 m). Moreover, the duration of the second vapour cavity formed at the valve is accurately simulated by the coupled DVCM using the VWS method. The efficiency of the VWS method is also verified in calculation of the piezometric head at the midpoint. Most unrealistic oscillations are avoided, and the shape is improved, but the main advantage consists in the very good timing accuracy.

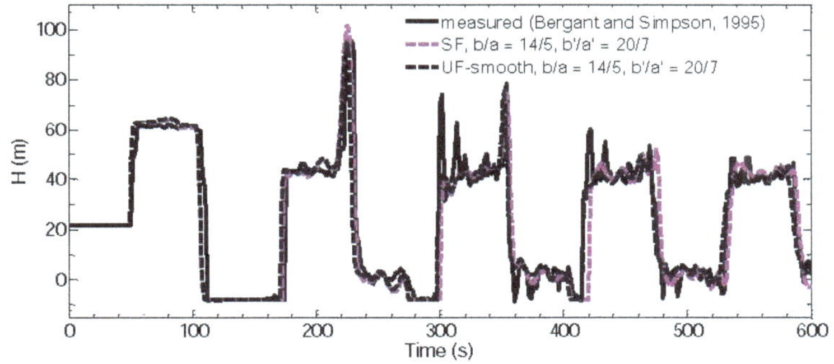

Fig. (5.27). Piezometric head at the valve showing efficiency of the VWS method.

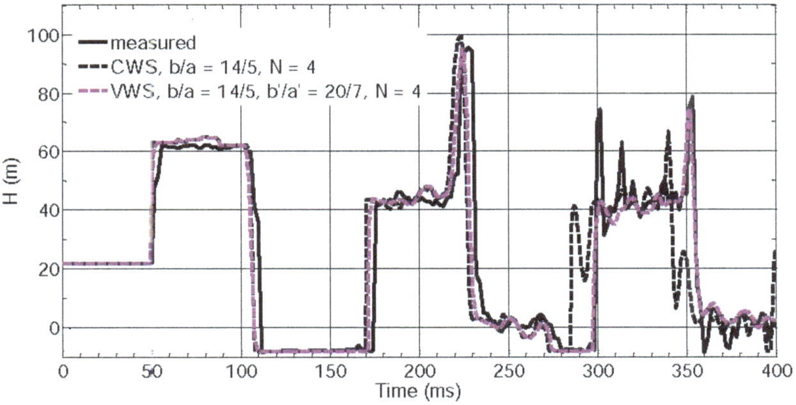

Fig. (5.28). Piezometric head at the valve obtained with UF in smooth pipe showing the preference of the VWS method to the CWS method.

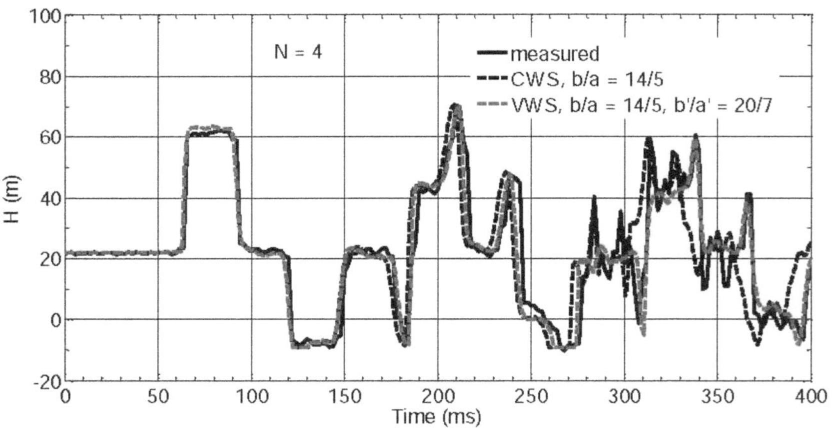

Fig. (5.29). Piezometric head at the midpoint obtained with UF in smooth pipe showing the preference of the VWS method to the CWS method.

Fluid Responses with Poisson and Junction Coupling

The experiment of Bergant and Simpson is used again to test the pressure evolution at the valve and at the midpoint in case of freely moving valve. This is the case of the simultaneous acting of Poisson and junction coupling. However, the above experiment does not precise whether the support material is quasi-rigid or viscoelastic.

The coupled DVCM is calculated by use of UF in smooth pipe, since this choice was already validated in the previous calculation. Moreover, a small number of reaches $N = 4$ is satisfying to obtain accurate results for all figures of this subsection. By taking a damping ratio ζ equal to 5 %, the equivalent stiffness of the system formed by the pipe and a unique anchor at the valve can be approximated [76]. According to the physical parameters of the experiment of Bergant and Simpson, the viscoelastic support structure has an equivalent stiffness approximated by $k = EA/L$ =391.97 kN.m^{-1} . A damping ratio of 2 % is assigned to the pipe while 3 % corresponds to the anchor which is modelled as linear spring.

Fig. (**5.30**) shows the piezometric head at the valve in case of quasi-rigid (QR) support and compares the results to that of the rigid support which corresponds to the no-junction coupling case. The equivalent viscous damping c is the same as the viscoelastic support. It is observed that the QR support solution converges to the fixed valve configuration (subsection 2.3.1) when the equivalent stiffness k (stiffness of pipe + support) is amplified compared to the viscoelastic support.

Hence, when the stiffness is so high, the viscous damping effect is no longer observed.

Fig. (**5.31**) shows the piezometric heads at the valve and at the middle of the pipe in case of viscoelastic support for a limited number of reaches ($N = 4$), and it compares these responses to those of the rigid support. Unlike the rigid support case which corresponds to the no-junction coupling, the freely moving valve with viscoelastic support does not provide accurate results. This inaccuracy can be due to the experimental conditions that may be carried out with rigid supports.

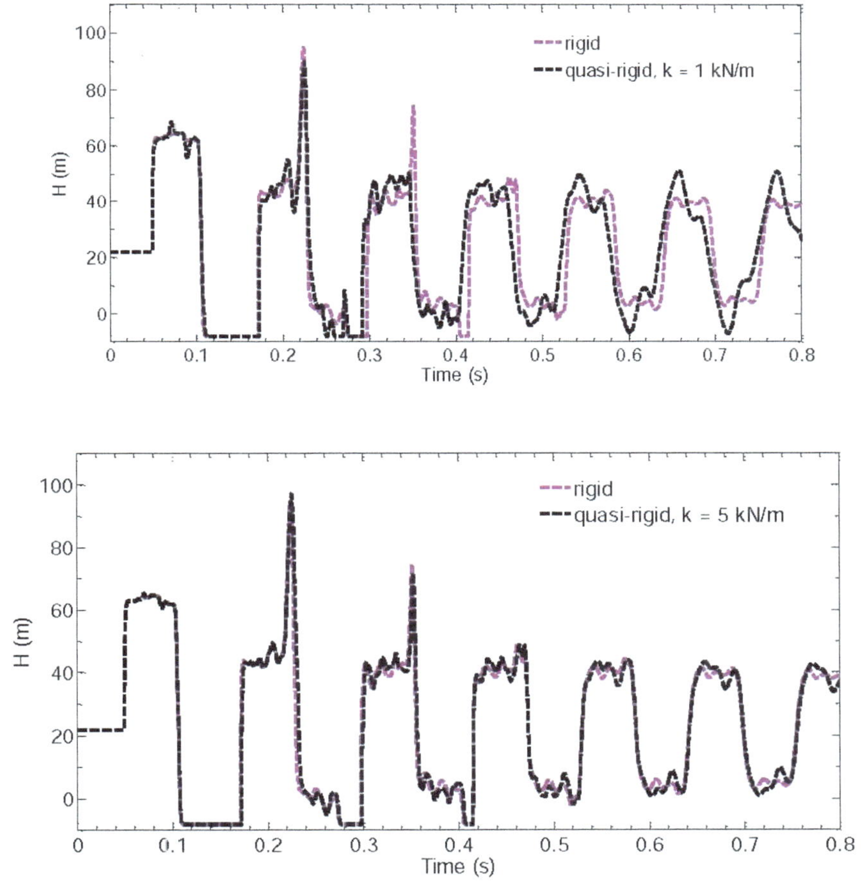

Fig. (5.30). Piezometric head at the valve and convergence of the QR solution to the rigid solution.

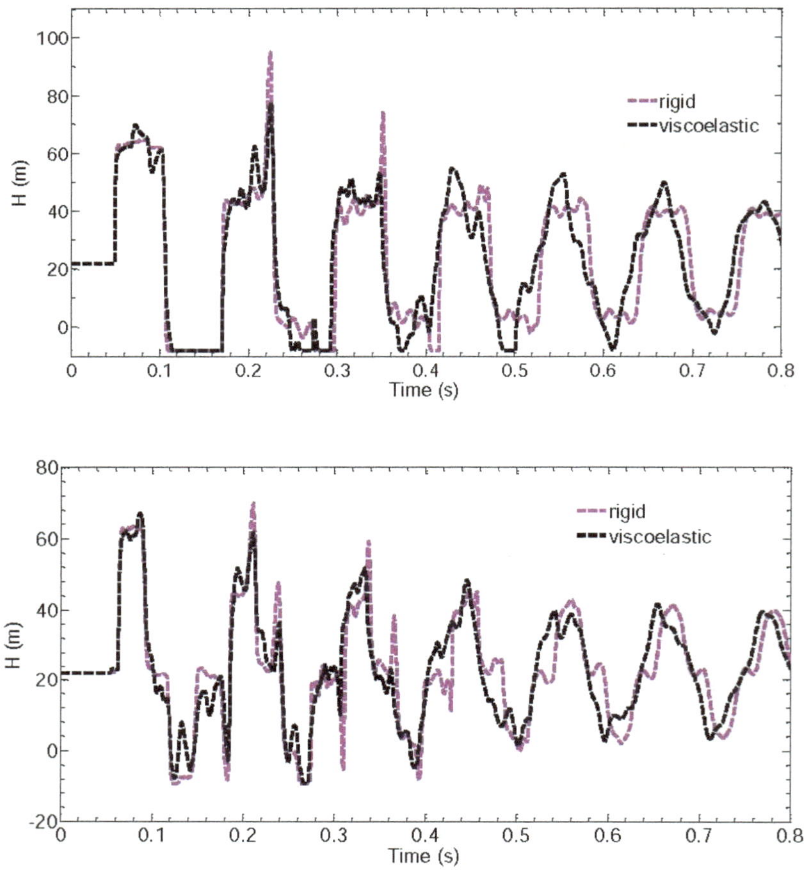

Fig. (5.31). Effect of viscoelastic support condition on the piezometric head. Top: at the valve; bottom: at the midpoint.

Structural Responses

This subsection deals with the structural responses along the pipeline, such as axial stress and axial velocity of the pipe. The experiment of Bergant and Simpson does not provide any information about anchors conditions and pipe stiffness, and unfortunately, there are no experimental results to compare the calculated structural responses.

Transient analysis of the liquid-filled piping systems has been developed by numerous researchers. The MOC was used to describe transients in piping systems for seven different wave components [84]. Lavooij and Tijsseling and later

Heinsbroek considered FSI in straight pipe and one-elbow pipe systems where the full MOC and MOC-FEM are compared [17 and 69]. The influence of support rigidity of pipelines on pressures and stresses was investigated in [85]. The authors analysed fluid and structure responses with respect to support rigidity and axial stiffness of 1 m of pipe. Their study dealt with the fact that pressures and stresses are as high as the supports are flexible. It has been established that Poisson coupling is important, if axial motion is dominant [69]. Based on the above research, it seems that accuracy of stress simulation depends on boundary conditions at the valve and whether the support rigidity and the stiffness of 1 m pipe are considered. Recently, flow-induced vibraticns of pipelines has been studied by considering the cavitating flow case [86].

For the rigid support case (fixed valve case) the boundary conditions imposed at the valve do not allow axial vibration. Thus, FSI is considered only *via* Poisson coupling and friction coupling; junction coupling is ignored since the valve is assumed to be rigidly supported. Nonetheless, according to previous research, if junction coupling is ignored, then the numerical errors are maximized. Fig. **(5.31)** displays the axial stress at the valve and compares the VWS method to the CWS method for the same number of reaches ($N = 4$) using UF in smooth pipe. The two solutions are identical from the beginning until the time $t = 110$ ms , which corresponds to a cavitating flow. After that, stress spikes of the CWS method becomes lesser than those of the VWS method, and in addition a timing delay can be displayed. The stress peak of the VWS method reaches 2.3 MPa at $t = 330$ ms , and it exceeds that of the CWS method by almost 0.5 MPa, which may be relevant in hydraulic engineering. It is worth noting that in case of fixed valve, axial stress at the valve is calculated with pipe equations unlike the free moving valve where boundary conditions impose proportional relation between axial stress and pressure. For this reason, the axial stress and the piezometric head histories at the fixed valve are not in phase. In addition, the two stress diagrams exhibit high-frequency oscillations, which are different from those of the piezometric head. This is due to the difference between the characteristic directions $\tilde{C}_f$ and $\tilde{C}_p$, along which the compatibility equations are integrated. The initial axial stress is obtained according to the initial conditions of the FSI model given in [2]. The calculation leads to 0.5 MPa (Fig. **5.32**). The Joukowsky pressure resulting from the valve closure represents the first exciting load that causes radial expansion of the pipe and an increase in the axial stress at the valve of about 0.44 MPa at $t = 50$ ms . This increase leads to a total stress of almost 0.94 MPa that slightly decreases to 0.8 MPa at $t = 70$ ms . The period of 20 ms corresponds exactly to the time occupied by the stress wave to travel from the valve to the reservoir and back to the valve.

The Joukowsky pressure maintained at the valve in the time interval between 50 ms and 100 ms keeps expansion of the pipe section adjacent to the valve. Subsequently, a positive axial stress slightly decreases therein. The reflected wave is attenuated from 0.94 MPa to 0.88 MPa. At $t = 70$ ms, this latter wave is superimposed to the previous stress of 0.8 MPa and leads to a first stress spike of almost 1.68 MPa. Two consecutive stress drops occur after this first spike. The first one follows the wave reflection at $t = 90$ ms and the second one, caused by the pressure drop starts at $t = 100$ ms. As a result, the axial stress falls to 0.2 MPa. The stress wave reflected from the reservoir comes to the valve at $t = 100$ ms, but the total stress does not exceed 0.55 MPa because of the vapour cavity formed at the valve. The effect of column separation on structure response does not stop at this moment, and the axial stress falls again to a negative value close to – 0.4 MPa. Similar analysis of FSI and column separation effect on the axial stress at the valve can be extended for the subsequent pulses.

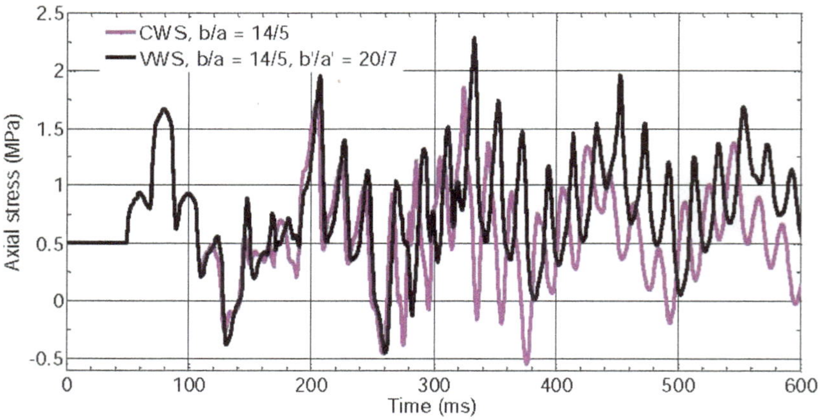

Fig. (5.32). Axial stress at the fixed valve in case of UF in smooth pipe and comparison between the CWS method and the VWS method.

For the viscoelastic support case which corresponds to the freely moving valve (junction coupling), the viscoelastic support and the quasi-rigid support are compared between each other. The equivalent stiffness of the quasi-rigid support structure is equal to 5000 kN.m^{-1}, whereas the calculated one of the viscoelastic structure supports is close to 391 kN.m^{-1}. Structural responses at the valve are displayed in Fig. (**5.33**) where the VWS method is used with- lesser number of reaches N = 4 and by use of UF in smooth pipe. It is shown that the cavity collapse at the valve involves axial stress peak greater than that of the quasi-rigid support. However, junction coupling reduces stress spikes at interior points like the middle of the pipe (Fig. **5.34**). Moreover, the viscoelastic support condition involves more displacement in the whole pipe. However, high-frequency responses involved by the rigid support condition are visibly reduced thanks to the viscoelastic consideration.

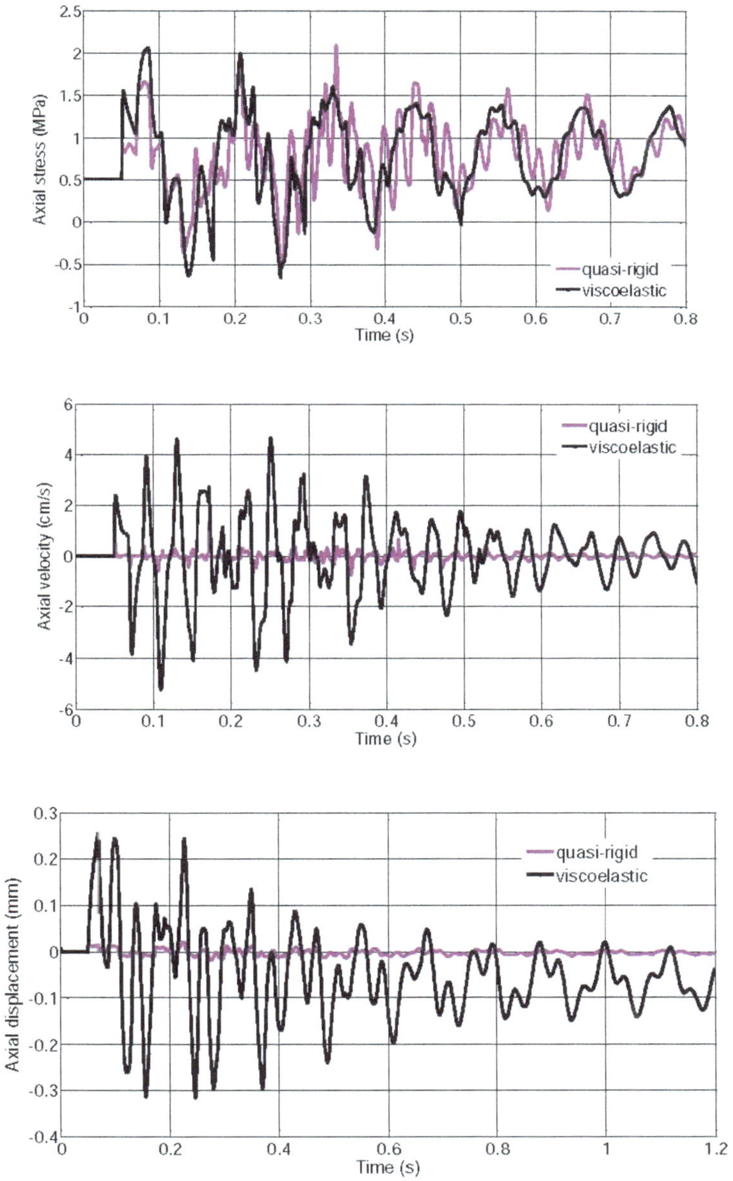

Fig. (5.33). Structural responses at the valve.

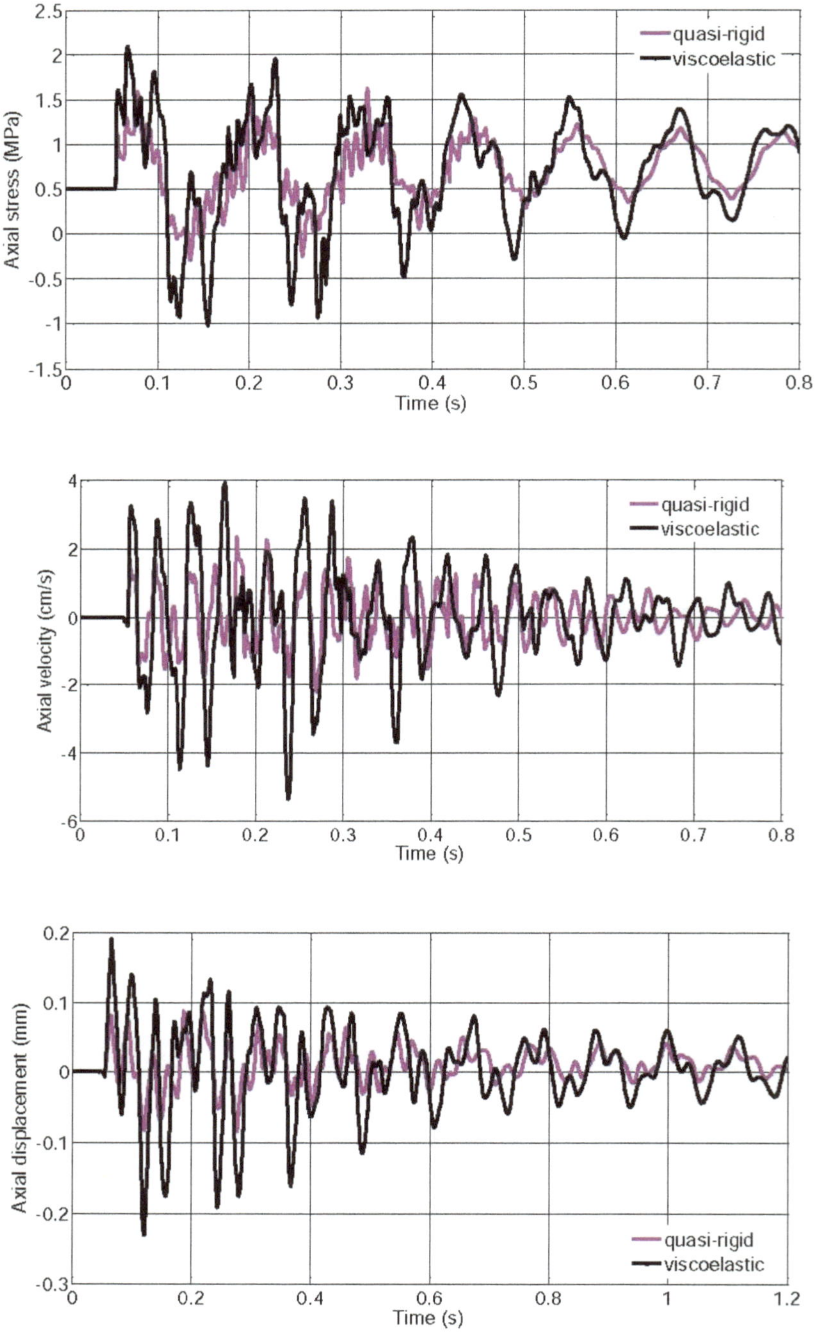

Fig. (5.34). Structural responses at the midpoint.

Simulation of the Coupled DGCM

The coupled DGCM proposed in chapter 4 is simulated and validated against experimental results of Bergant and Simpson (1995) in the low-velocity case considered previously. The fluid responses are briefly presented and compared against the coupled DVCM. Also, some precious techniques obtained from the previous investigation of the coupled DVCM are applied to the coupled DGCM, such as the use of SG instead of the RG and the preference of UF in smooth pipe to the rough pipe. Moreover, the conclusions reported for the coupled DVCM in case of vibrating pipe and the effect of its support material are not detailed in this subsection because it was observed that this assumption does not give accurate results for the present experiment. The valve is assumed to be fixed and only Poisson and friction coupling matter.

The calculation of the coupled DGCM is performed for two different values of the weighting factor ψ (0.5 and 1). Fig. (**5.35**) shows the piezometric head at the valve and at the midpoint and compares the coupled DGCM to the classical DGCM and the coupled DVCM for $\psi = 0.5$, while Fig. (**5.36**) displays the same comparison for $\psi = 1$. It appears from this comparison that the weighting factor variation is of high impact on the classical solution, however, the coupled solution is slightly influenced with it. If $\psi = 1$, the magnitude of the coupled solution is improved compared to that of $\psi = 0.5$. Moreover, the piezometric head obtained with the coupled DGCM is more accurate than that of the classical one regardless the value of ψ. The solution obtained by the coupled DGCM is then compared against the solution of the coupled DVCM for $\psi = 1$ (Fig. **5.37**). Although the timing accuracy exhibited by this latter, some pressure pulses cannot be displayed with it, and in contrast, they are nicely simulated by the coupled DGCM.

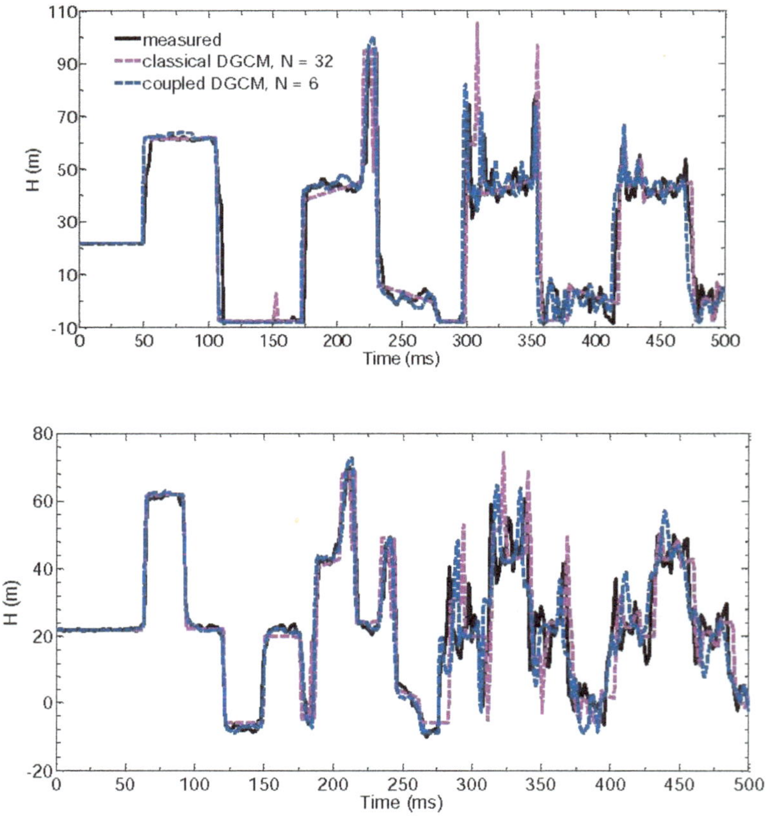

Fig. (5.35). Piezometric head histories with UF in smooth pipe for $\psi = 0.5$. Top: at the valve; bottom: at the midpoint.

The coupled DGCM is used in next step to predict structural responses along the pipeline where UF in smooth pipe is considered with a weighting factor $\psi = 1$. The axial stress at the valve, the axial stress and the axial displacement at the midpoint are simulated in Fig. **(5.38)**. The axial displacement at the valve is equal to zero during the transient according to the imposed boundary conditions where the valve is assumed to be rigidly fixed (rigid support). (Fig. **5.38**) compares these structural responses against those obtained by the coupled DVCM. It is observed that axial stress diagrams have the same frequencies, however, the solution of the coupled DGCM is characterized by a dominant compression stress at the valve and at the midpoint. For instance, the stress at the valve reaches -0.7 MPa at time $t = 377$ ms by calculation of the coupled DGCM, while the coupled DVCM gives a stress drop of -0.45 MPa at time $t = 260$ ms. Except the initial stress, the axial stress predicted by the coupled DGCM is lower than that of the coupled DVCM regardless the time and the location of the computational section. Subsequently, the buckling mechanism is assumed to be larger when the coupled DGCM is used for column separation prediction.

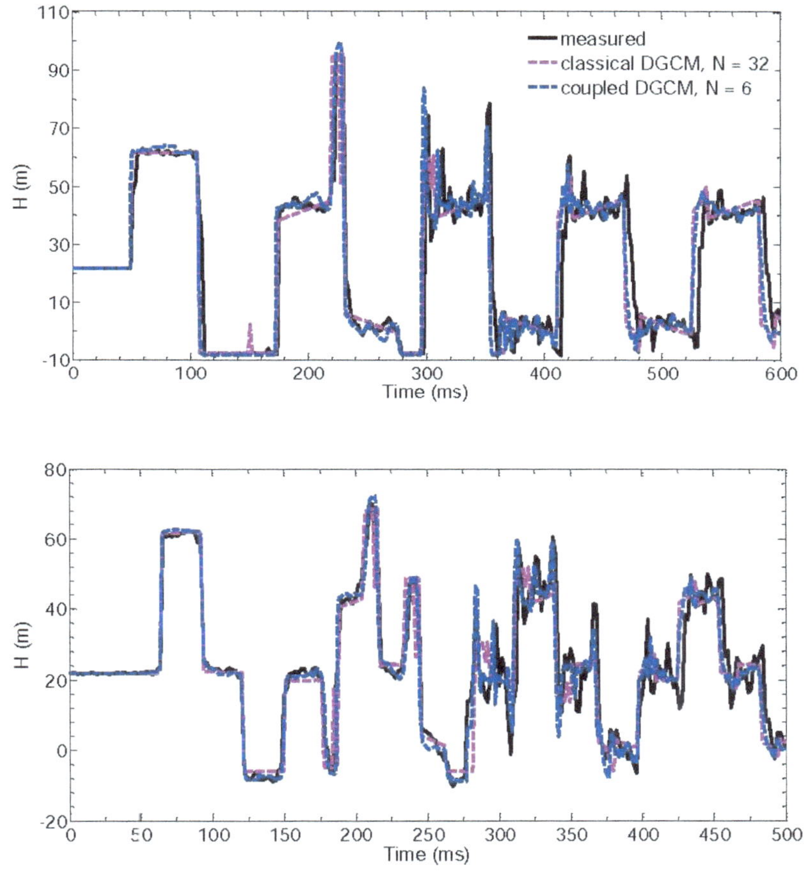

Fig. (5.36). Piezometric head histories with UF in smooth pipe for $\psi = 1$. Top: at the valve; bottom: at the midpoint.

The axial displacement of the pipeline matters also in mechanical design. The prediction of the axial displacement at the valve is shown in Fig. (**5.38**) where the coupled DGCM is compared against the coupled DVCM. The stress diagrams show strong oscillations characterizing the coupled DGCM up to $t = 250$ ms especially at the midpoint. According to the pressure diagrams of Fig. (**5.37**) that shows very good agreement with the experiment provided by the coupled DGCM, it can be assumed that axial stresses simulated by this latter are more accurate than those of the coupled DVCM. Axial displacement curves are very similar until $t = 120$ ms (opening of the cavity). Up to this time, some discrepancies can be observed in magnitudes, but lesser than those exhibited by stress curves. At $t = 270$ ms, the coupled DVCM exhibits strong drop that reaches almost -0.11 mm, so that it exceeds the coupled DGCM solution by almost 0.65 mm. These discrepancies can be due to the longer time of cavity existence characterizing the coupled DGCM as shown in the above pressure diagrams, since the gas dissolving takes a longer time than vapour condensation [6].

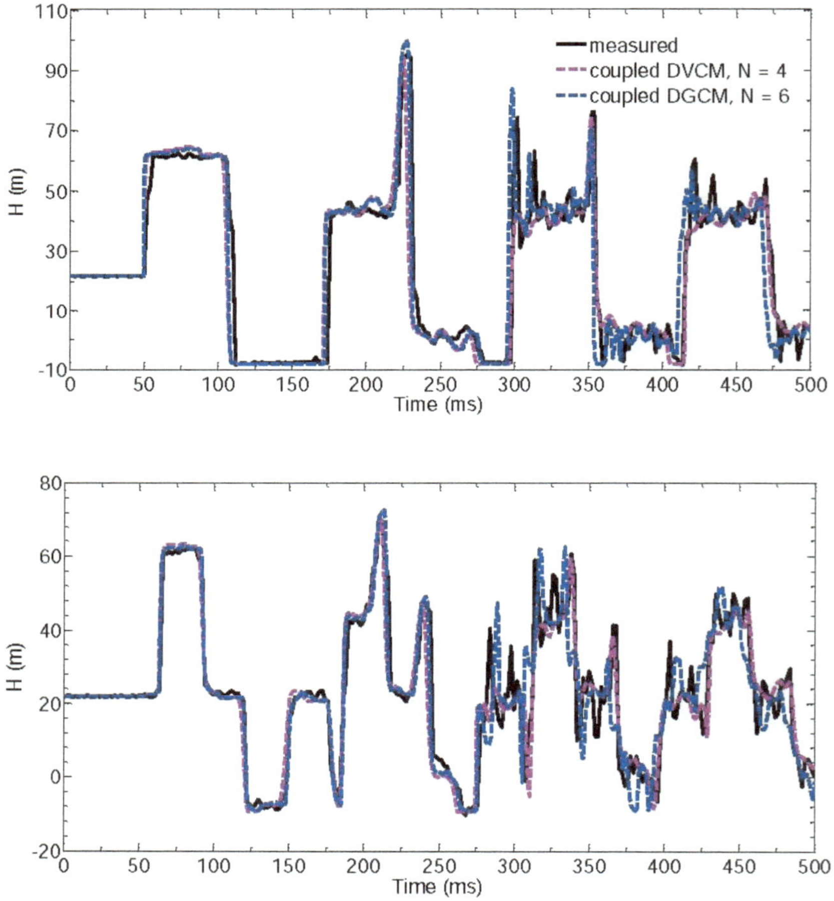

Fig. (5.37). Piezometric head histories and Comparison between the coupled DGCM and the coupled DVCM for $\psi = 1$. Top: at the valve; bottom: at the midpoint.

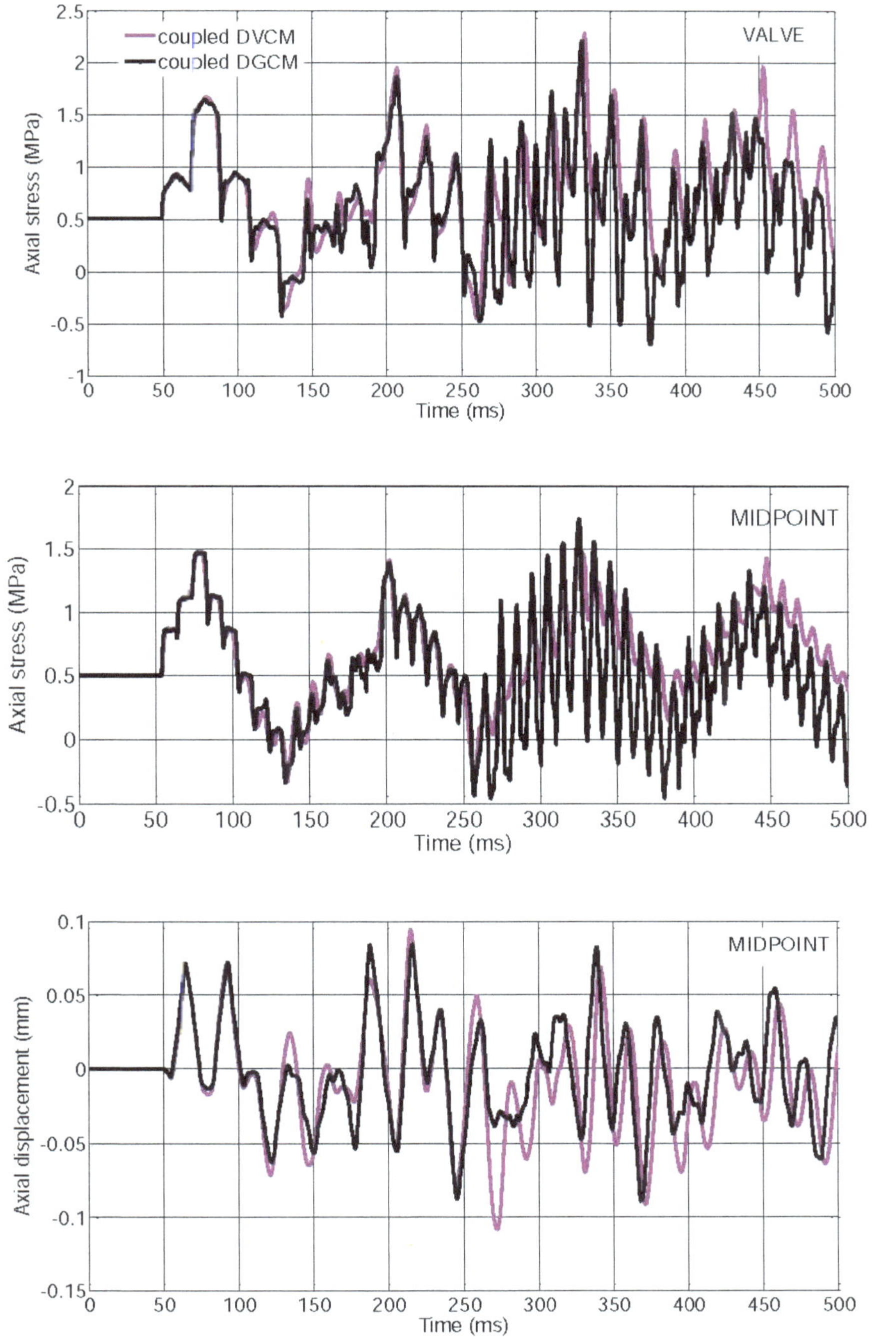

Fig. (5.38). Structural responses at the valve and the midpoint.

PASSIVE COLUMN SEPARATION FLOW REGIME

This section deals with the prediction of column separation without pressure pulse, called "passive column separation flow regime" [82]. This cavitation corresponds to the high velocity case involves column separation event with a maximum pressure lesser than the Joukowsky valve closure pressure. Pressure surges following the cavity collapses no longer occur. The reflected pressure rise and the collapse rise are incorporated and form a unique pressure pulse. The proposed models are validated against experimental results in this subsection. The experiment of Bergant and Simpson described in chapter 1 is used again to validate the proposed models. The computations are performed for a rapid-closure downstream valve and an upward sloping pipe. The computational parameters noted in subsection 2.1 are still valid for this section except the high-velocity case which corresponds to $V_0 = 1.4\ \text{m.s}^{-1}$.

Validation of the Coupled DVCM

Figs. (**5.39** and **5.40**) show the simulation of the piezometric head, respectively, at the fixed valve and the middle of the pipe using the classical DVCM for $N = 16$. The VWS method is used in calculation. Unlike the low-velocity case, the smooth pipe solution does not agree with the experimental result unless for the first and the second pressure rise. In the third and the fourth ones, the damping effect is amplified, and the timing error becomes greater. Both rough pipe solution and SF solution are better than the smooth pipe one, but they still involve shape errors. Despite that, the SF solution has the advantage of the short computation time.

Piezometric heads at the valve and at the middle of the pipe are also predicted using the coupled DVCM for lesser number of reaches ($N = 8$). The VWS method is used again according to Table **5.1**, but different integers a and b are used; $b/a = 20/7$ and $b'/a' = 23/8$. This choice was based on different tests performed for several combinations of integers. Figs. (**5.41** and **5.42**) show the piezometric heads in case of fixed valve, so that junction coupling is ignored.

It is observed that the use of UF in rough pipe leads to more accuracy than the use of SF solution, which is better than the result provided by use of UF in smooth pipe. This latter solution involves non-realistic damping accompanied with large timing delay as mentioned above for the classical DVCM. This phenomenon is attributed to the inadequate selection of the weighting function [25]. Vardy and Brown (2003, 2004) gave Common features for smooth and rough pipes was given in [87 and 88], namely the linearly dependent viscosity at the wall ν_w, the uniform viscosity in the core ν_c and the constant values in time of these two viscosities.

In the present study, ν_w is assumed to be equal to ν_c (equal to the laminar viscosity ν_{lam}). This assumption leads to acceptable results which are better than those of the rough pipe in the low-velocity case but not in the high-velocity case. The simulation shows that the unsteady part used for smooth pipe is greater than that of the rough pipe. This may be due to the existence of diphasic regions in some reaches of the pipeline during transient flow leading to important variation in the kinematic viscosities. Otherwise stated, distributed cavitation may occur along the pipeline, because of the precursor wave (Poisson-coupling induced cavitation) that follows from pressure drop waves caused by the wall radial expansion [2]. The precursor waves result in the opening and the collapse of multi-vaporous cavities at the midpoint at high frequency and low magnitude.

Although discrepancies involved by the smooth pipe solution, the use of the VWS method for the coupled DVCM in high-velocity case can simulate column separation better than the classical DVCM. Additionally, the coupled solution converges more rapidly and involves more stability. Nevertheless, the computation time of the coupled DVCM is increased compared to the classical DVCM, especially if UF is used.

In Figs. (**5.43** and **5.44**), piezometric heads at the valve and at the midpoint are displayed in case of freely moving valve. The rough-pipe assumption is considered for UF since it is validated for the previous case (Figs. **5.41** and **5.42**). Junction coupling is considered for such configuration in which the support material matters. The QR support solution is compared against experimental records and against the viscoelastic support solution. The equivalent stiffness of the structure is equal to $5000\ \text{kN.m}^{-1}$. The computation shows that this latter fails in simulation of passive column separation. This discrepancy may be due to the real support material used in the experiment of Bergant and Simpson.

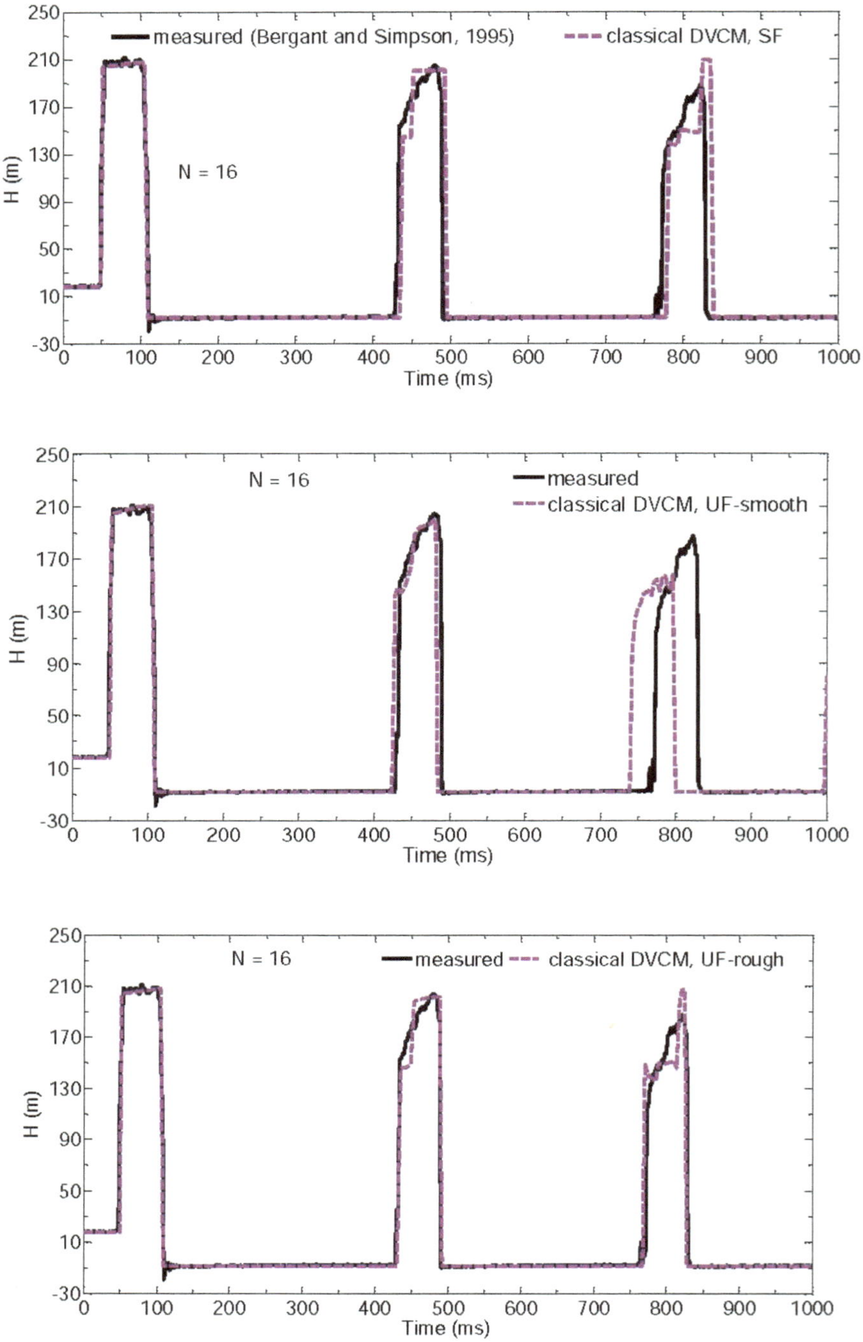

Fig. (5.39). Piezometric head at the valve given by the classical DVCM.

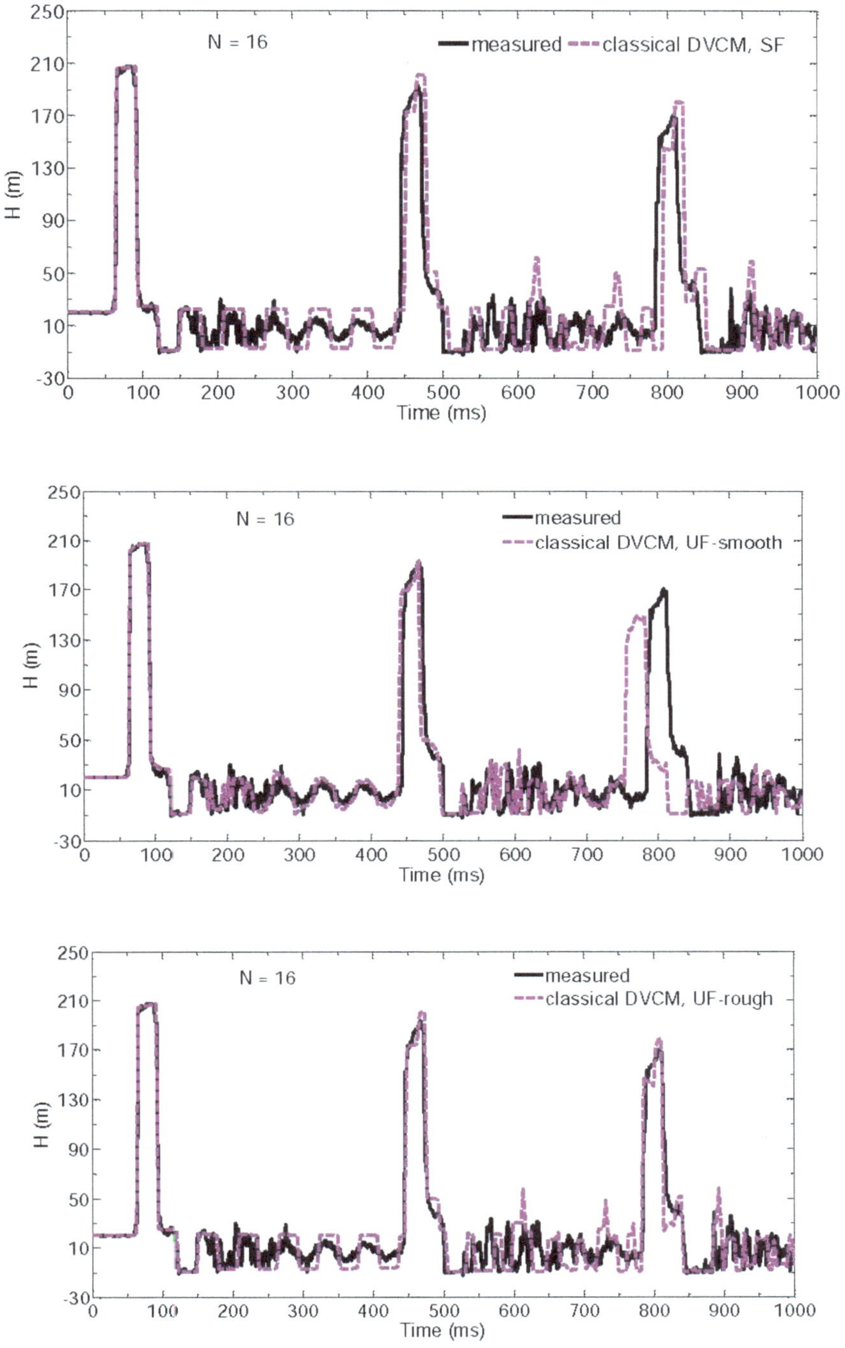

Fig. (5.40). Piezometric head at the midpoint given by the classical DVCM.

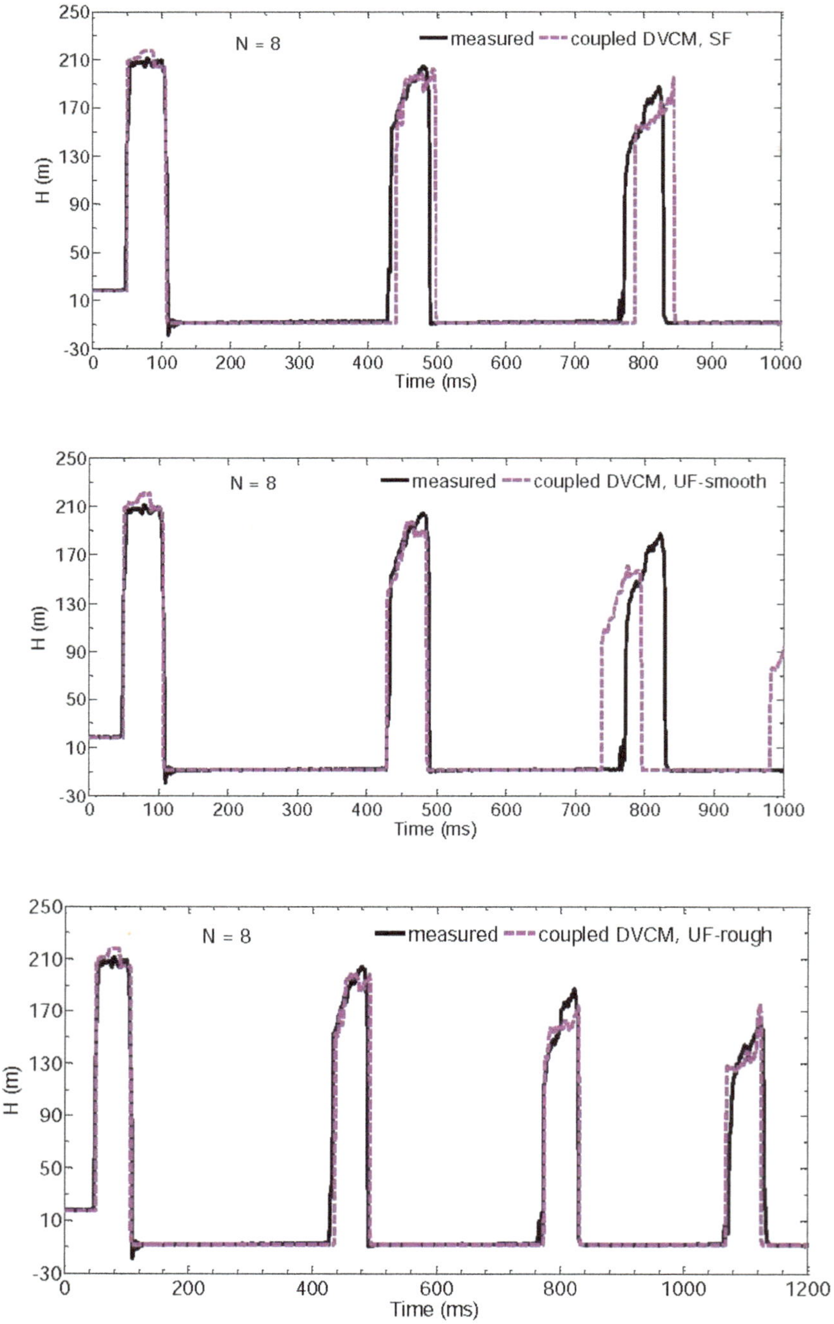

Fig. (5.41). Piezometric head at the valve given by the coupled DVCM with fixed valve.

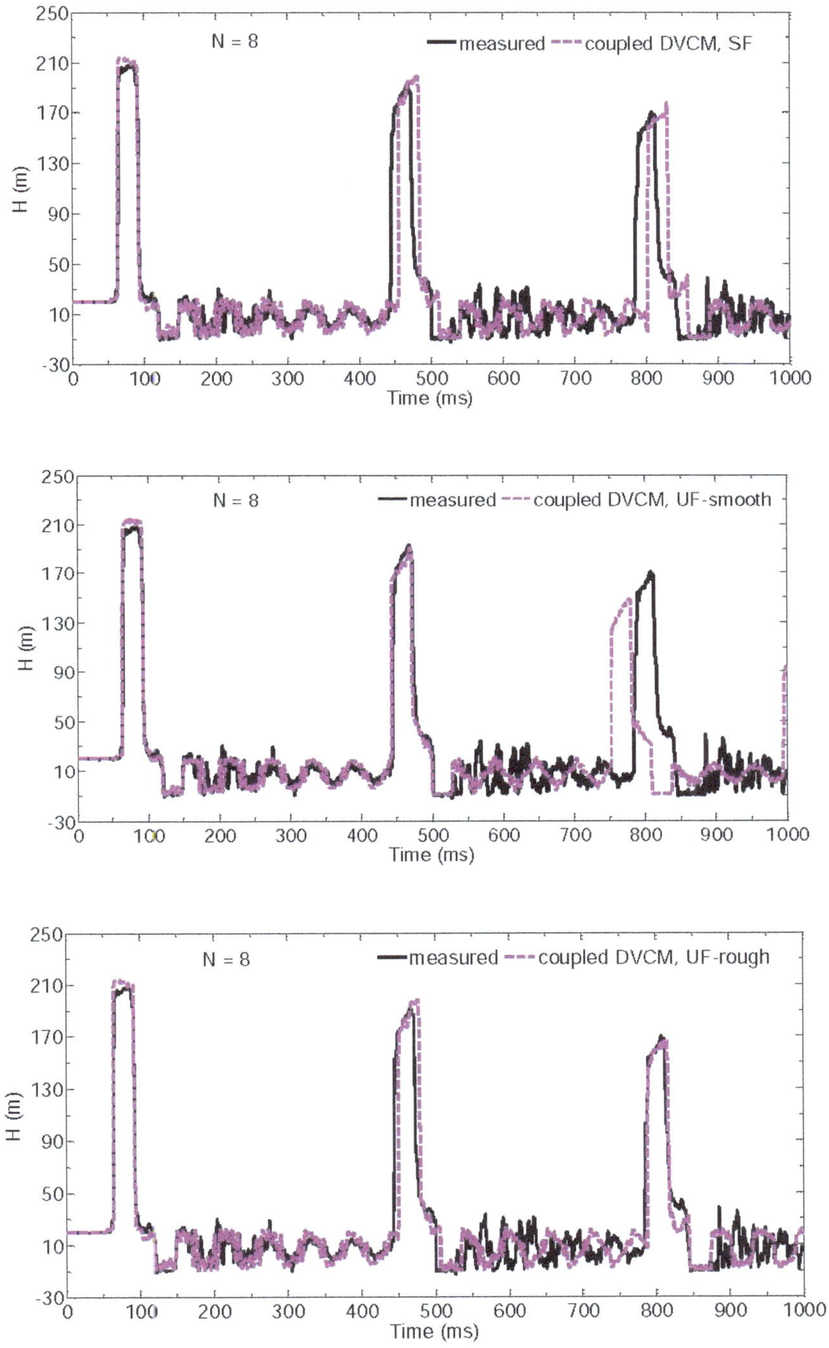

Fig. (5.42). Piezometric head at the midpoint given by the coupled DVCM for fixed valve.

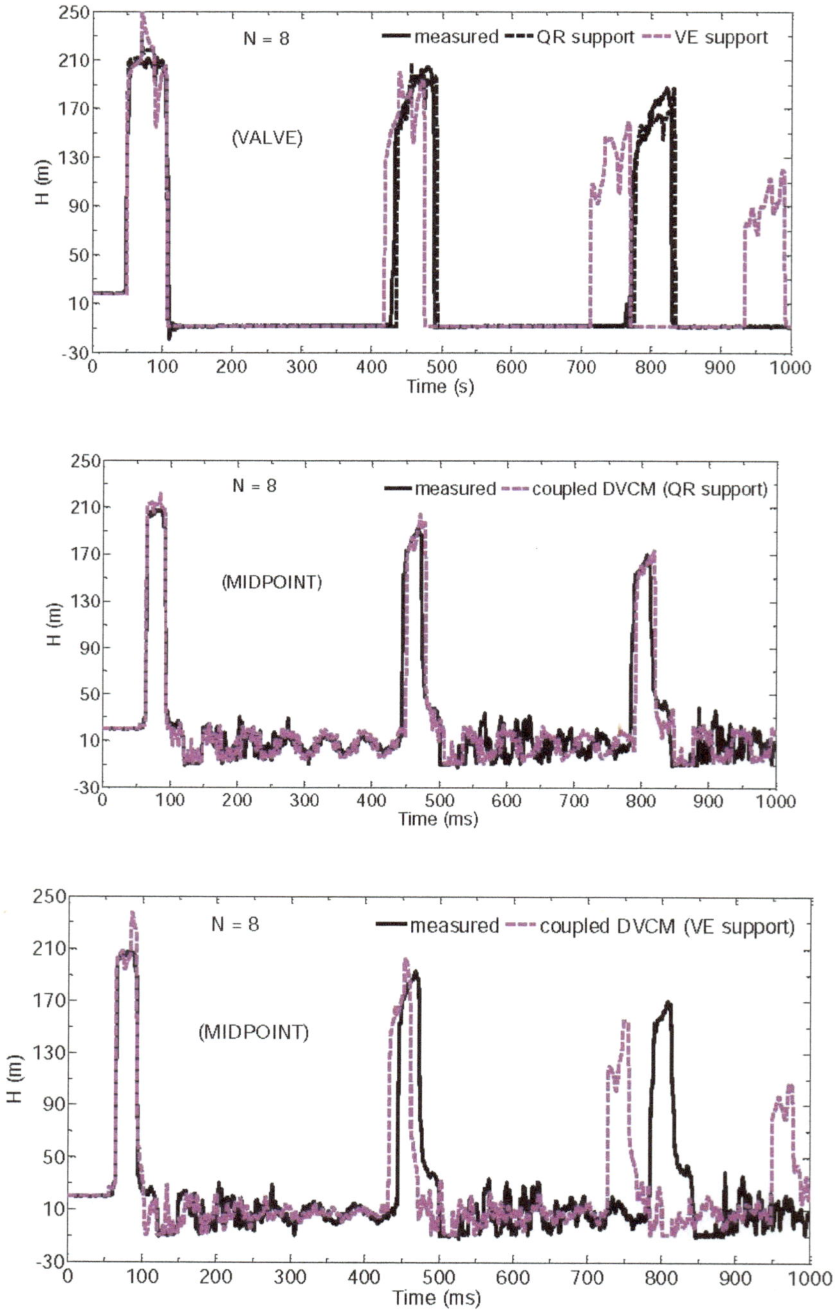

Fig. (5.43). Piezometric heads provided by the coupled DVCM for free valve.

Validation of the coupled DGCM

Since the classical DGCM and the coupled DGCM are considered as efficient alternative for the classical DVCM and the coupled DVCM, passive column separation flow regime is attempted to be accurately predicted with them. However, Figs. (**5.44** and **5.45**) show some discrepancies involved by the classical DGCM, despite the high number of reaches used for calculation. Whatever the type of friction coupling, a timing delay is observed at the valve and at the midpoint. Moreover, unrealistic oscillations appear at the fourth pressure spike, but the magnitudes of the pressure head are in acceptable agreement with experiment.

Figs. (**5.46** and **5.47**) illustrate the preference of the coupled DGCM to the classical DGCM in case of fixed valve. The use of UF in smooth pipe leads to more accuracy regarding timing whereas UF in rough pipe seems more efficient in simulation of shape and friction damping.

Fig. (**5.48**) displays the influence of junction coupling on the numerical solution in case of UF in rough pipe. Two types of pipe support are compared, namely the QR support and the viscoelastic support. For the former case, the equivalent stiffness is assumed to be very high and equal to $5000\ \mathrm{kN.m^{-1}}$ as defined for the coupled DVCM. The rational number of the WSA scheme is $b/a = 20/7$; this choice was based on different tests performed for several combinations of integers, and it was found that it is better than $23/8$ and $17/5$. The most important result that can be focused on (Fig. **5.48**) consists in the similarity of the two solutions and the good agreement of both with the experimental result. However, some unrealistic oscillations are still involved by the viscoelastic support solution. (Figs. **5.49** and **5.50**) exhibit comparison between the QR solutions provided by the coupled DVCM and the coupled DGCM in case of UF in rough pipe. The simulation of the piezometric heads at the free valve and the midstream of the pipe shows slight preference of the former. This can be due to the use of the VWS method which takes account of the variation of the pressure wave-speed in the liquid. Hence, the unique weakness of the coupled DGCM is the ignorance of distributed cavitation in the reaches.

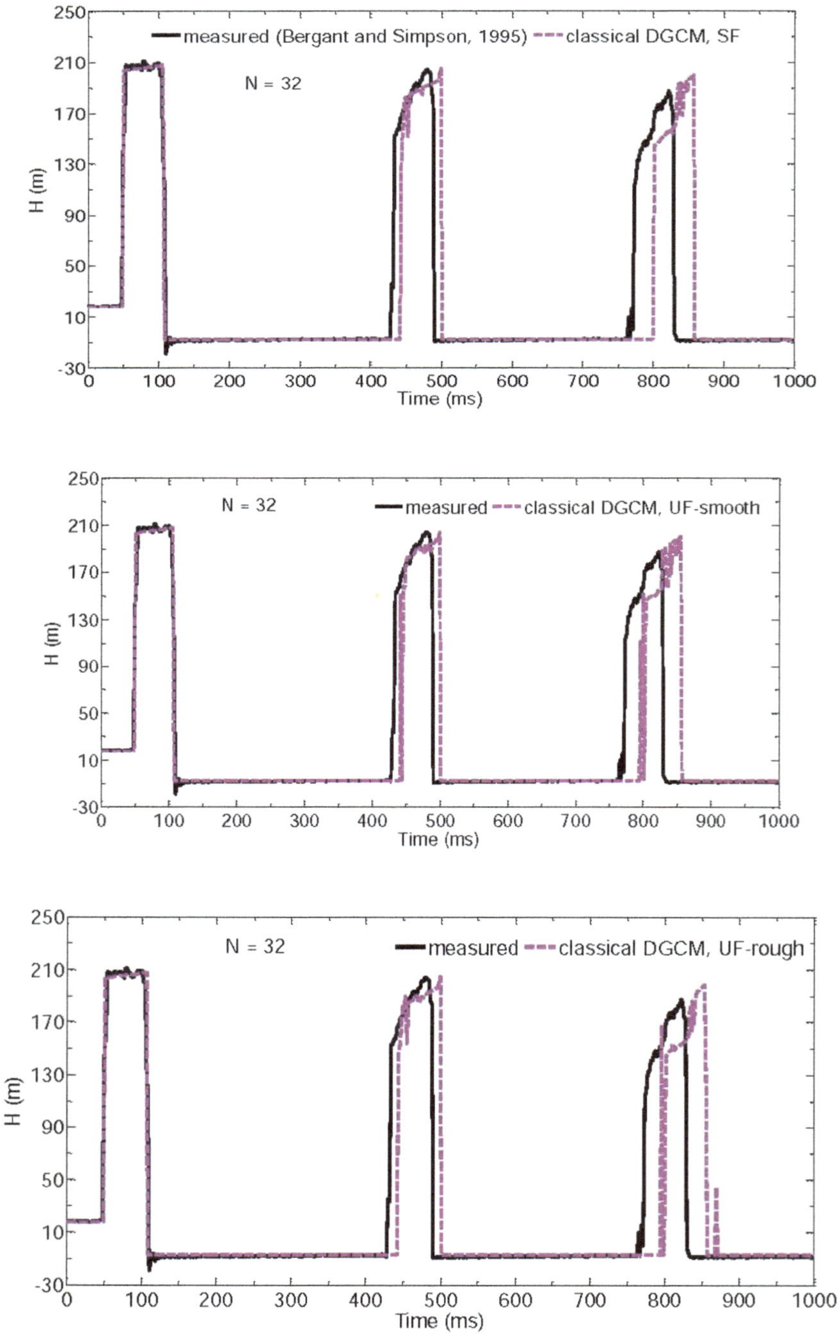

Fig. (5.44). Piezometric head at the valve provided by the classical DGCM.

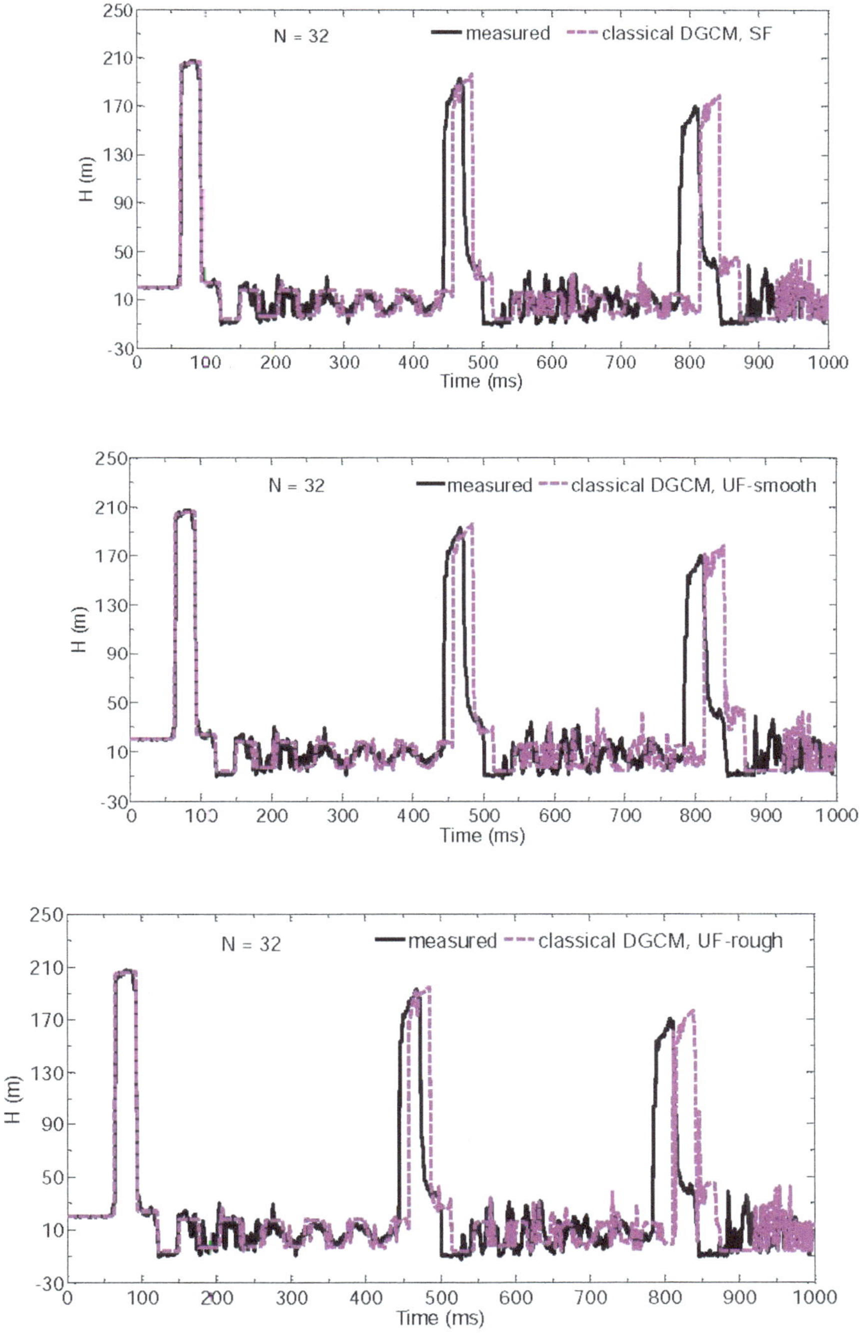

Fig. (5.45). Piezometric head at the midpoint provided by the classical DGCM.

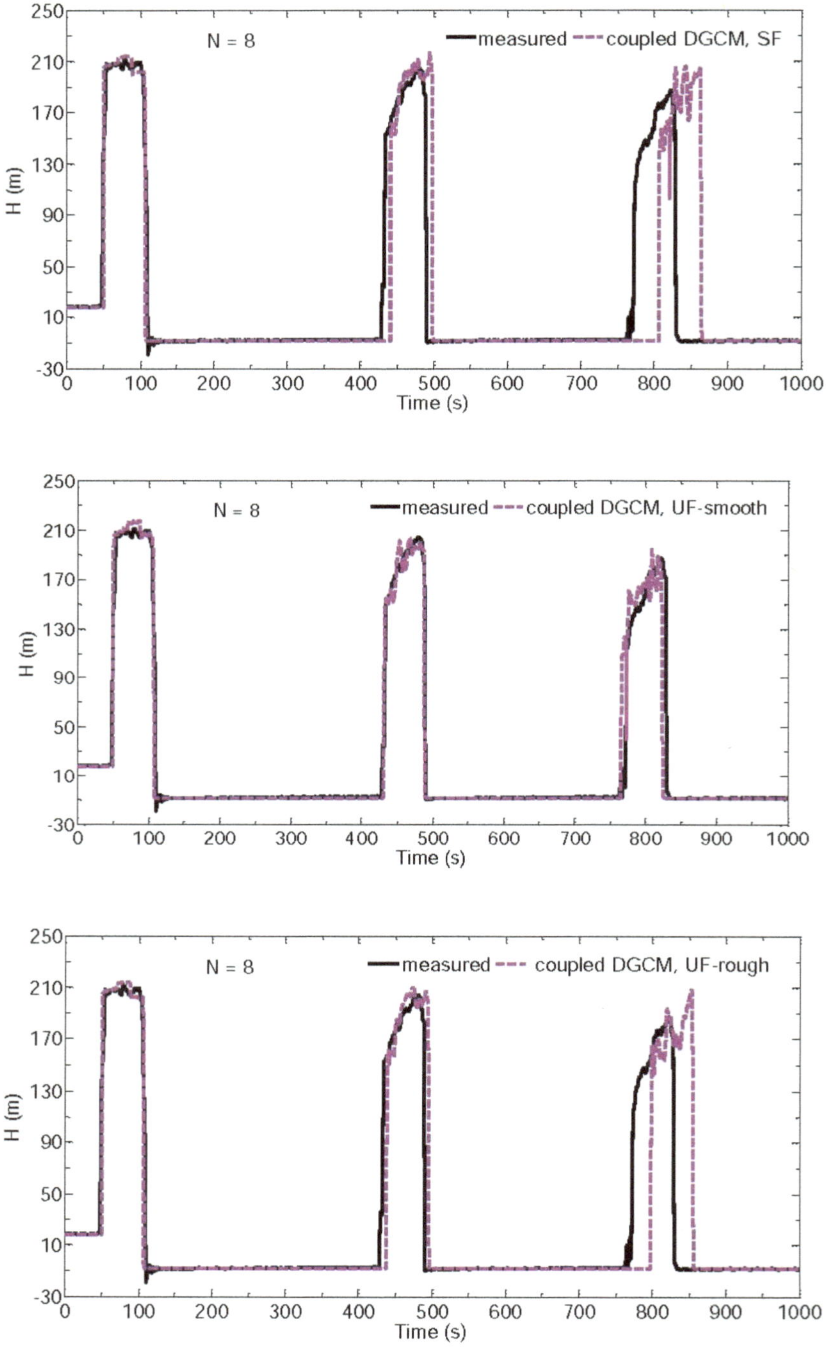

Fig. (5.46). Piezometric head at the fixed valve provided by the coupled DGCM.

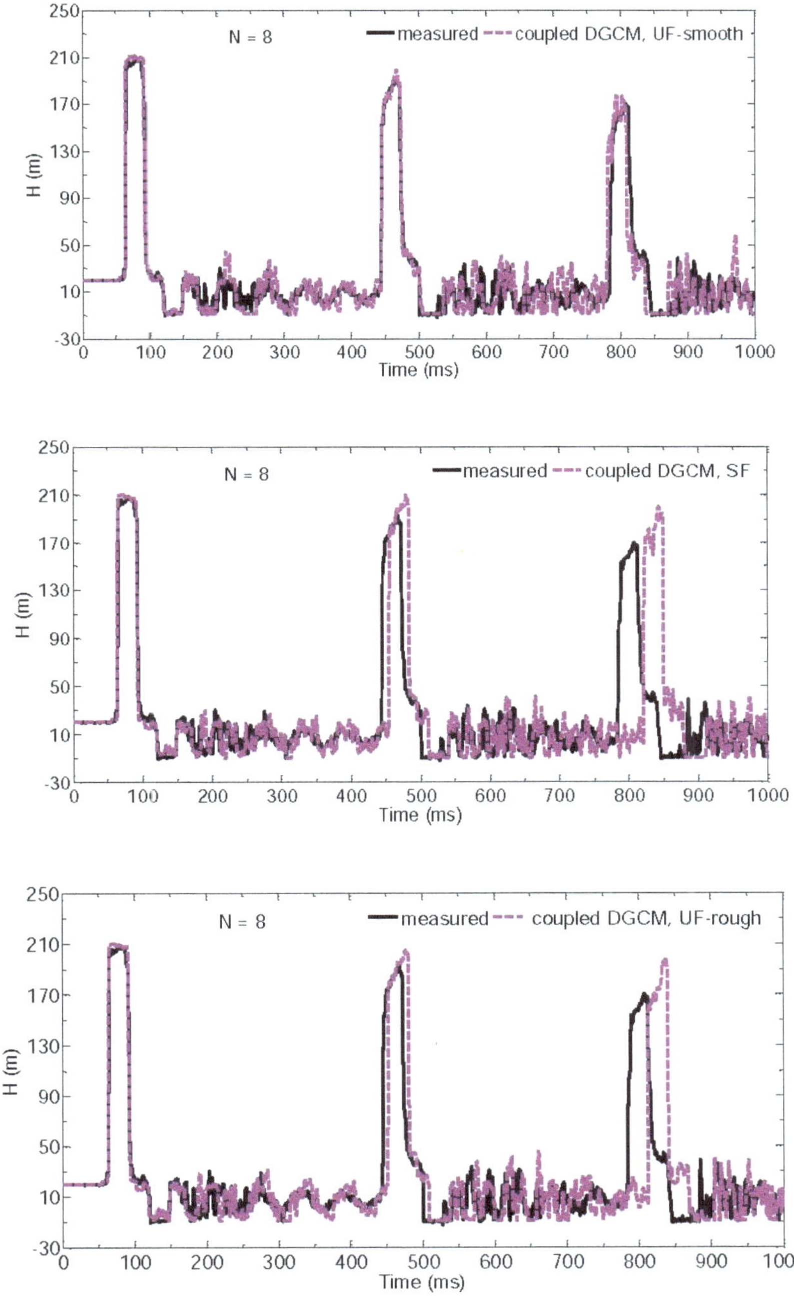

Fig. (5.47). Piezometric head at the midpoint provided by the coupled DGCM (fixed valve).

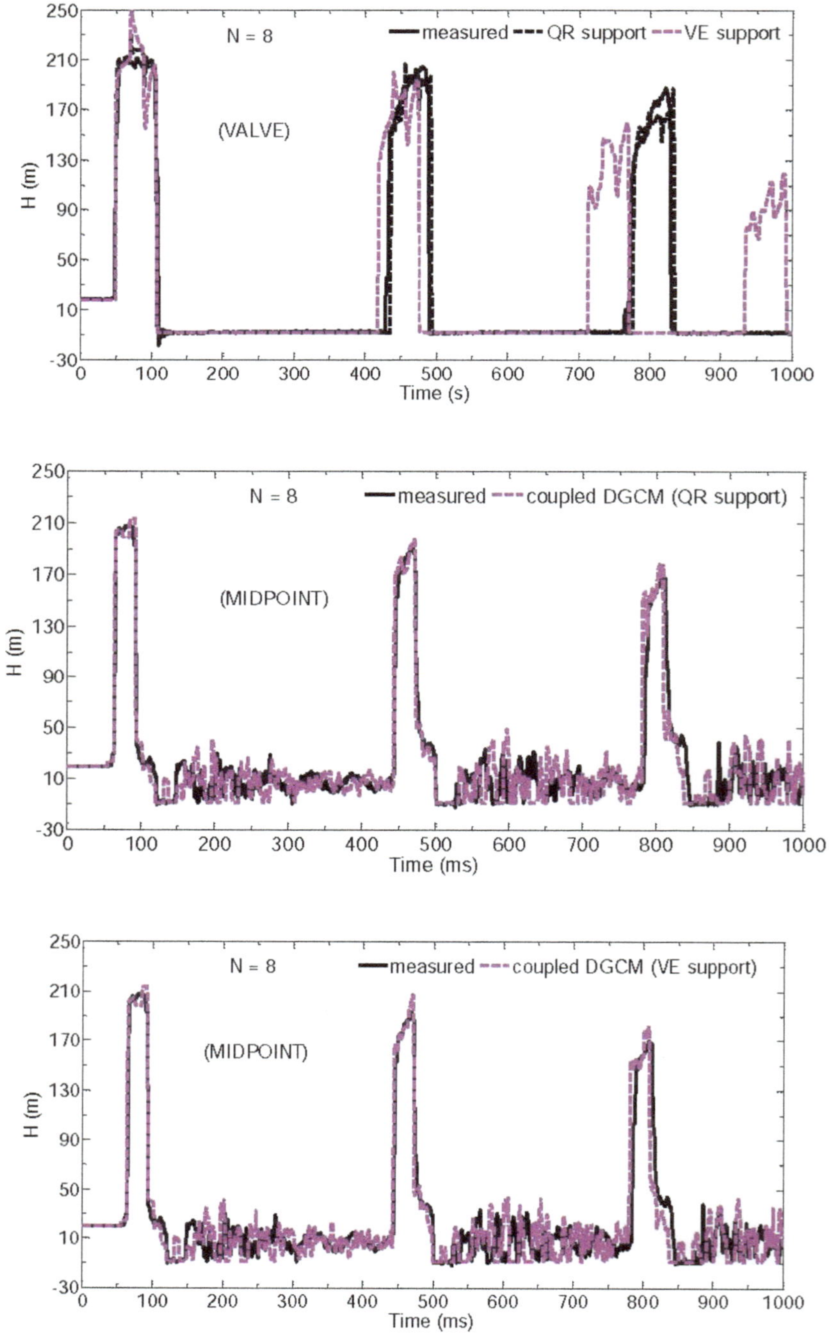

Fig. (5.48). Effect of the support material on the piezometric head in case of free valve.

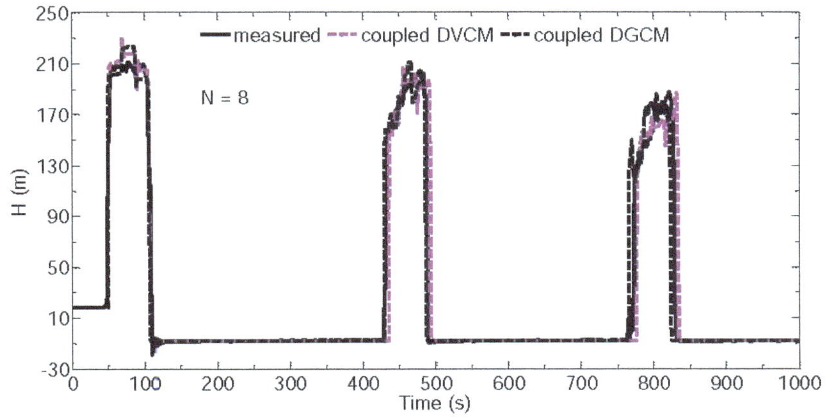

Fig. (5.49). Piezometric head at the free valve with QR support and comparison between the coupled DGCM and the coupled DVCM with UF in rough pipe.

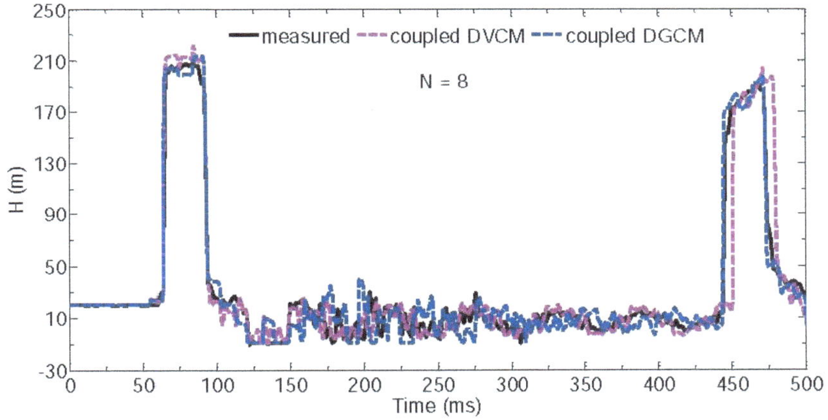

Fig. (5.50). Piezometric head at the midpoint in case of free valve with QR support and comparison between the coupled DGCM and the coupled DVCM with UF in rough pipe.

CONCLUSION

This chapter dealt with the calculation of fluid transients in elastic pipelines with emphasis on the proposed models that consider FSI in their governing equations. Water hammer without cavitation has been studied in the first step in order to reveal the importance of the WSA scheme of the MOC in simulation. The Benchmark problem A (BPA) has been used to analyse the numerical results. The effect of

Poisson coupling has been carried out and the precursor wave effect has been detailed. Also the effect of junction coupling on the solution has been described.

The cavitation case has been studied in the second step by considering only localized cavitation referred to as liquid column separation. Two types of column separation has been simulated using classical and coupled models (proposed models). The first is the active column separation flow regime which coresponds to the low initial velocity case characterized by short duration pressure pulse. The second is the passive column separation flow regime in which the pressure pulse does not exceed the Joukowsky pressure pulse. In the two cases, the calculations have been carried out using four types of column separation models. Two among them are available in the literature: the classical DVCM and the classical DGCM. And the others are proposed by the present author, namely, the coupled DVCM and the coupled DGCM. Moreover, three types of friction coupling have been used: the steady friction (SF), the unsteady friction (UF) in smooth pipe and the UF in rough pipe. Also, junction coupling effect has been investigated by considering fixed and freely moving valves. In the whole study, the coupled models have been validated against experimental results, and they are then preferred to the classical models regarding accuracy and computational efficiency. The classical DGCM is preferred to the classical DVCM for column separation calculation, however, it does not allow prediction of structural responses, such as axial stress and axial displacement. The coupled DVCM is preferred to the classical DVCM and the classical DGCM but it does not take account of gas release phenomenon. UF damping have been accurately simulated. The smooth pipe assumption is preferred for active column separation whereas the rough pipe assumption is more accurate for passive column separation. The effect of junction coupling is better in this second type of column separation. The QR support is better than the viscoelastic support especially when the coupled DVCM is used. This model is slightly better than the coupled DGCM, thanks to the variable wave-speed (VWS) method.

CHAPTER 6

Numerical Results for Viscoelastic Pipelines

Abstract: This chapter deals with the numerical results obtained from simulation of fluid transients in viscoelastic pipes. For the non-cavitating flow case, the simulation shows that the pressure fluctuations are rapidly attenuated, and the overall transient pressure wave is delayed in time due to the retarded deformation of the pipe-wall. In addition, the simulation of the cavitating flow gives reasonable results compared to experimental data.

Keywords: Circonferential strain; Creep-compliance; Experimental result; Viscoelastic pipelines.

INTRODUCTION

The viscoelastic behaviour of the polyethylene (PE) pipe determines the pressure response during transient. The calibrated creep functions provided by Covas in [1] are used in simulation of water hammer without cavitation. The experimental data of Stoinov and Covas are used to validate calculations of the classical models as well as the proposed models without cavitation.

Case Studies

Case 1: Water Hammer in HDPE Pipeline

The experiment of Stoinov and Covas carried out in 277 m HDPE pipeline, 63 mm diameter, at Imperial College (IC) is mainly used to validate numerical calculations. The experimental facility is presented in Subsection 6.4 of Chapter 1. The pipeline was fixed to a concrete structure with plastic brackets, 1 m spaced, and with metal frames at the elbows, to avoid any axial movement. This facility was particularly designed and constructed for the investigation of novel leak detection techniques based on the generation of transient events in a water pipe system. Collected transient data were essentially used for calibration and validation of the Hydraulic Transient Solver (HTS) and Inverse Transient Analysis (ITA) proposed by Covas in [1].

Transient pressure and circumferential strain data were collected at several sites. The effect of pipe-wall viscoelasticity was observed in specific features of the

Abdelaziz Ghodhbani, Ezzeddine Haj Taïeb, Mohsen Akrout & Sami Elaoud

transient pressure signal (*i.e.*, initial pressure peak and major energy dissipation) and in stress-strain curves during the occurrence of transient events (*i.e.* mechanical hysteresis). Viscoelastic behaviour is typically described by a creep-compliance function, $J(t)$. Results of the experimental tests were discussed, and final remarks were made on how to characterise the viscoelastic behaviour of PE as a material and when it is integrated in a water pipe system.

In 2003, Covas gave an extensive work in which she analysed the effects of several parameters on fluid and pipe responses: steady friction (SF), unsteady friction (UF), boundary conditions and viscoelasticity [1]. The emphasis was made on the major effect of viscoelastic behaviour on pressure and strain damping and dispersion. To obtain accurate results, calibrations of the transient solver was used. However, in the present study, any of these calibrations will be adopted; some relevant parameters mentioned in [1] will be useful, such as the ignorance of the UF effect (frictionless modelling), the number of Kelvin-Voight elements and the pressure wave-speed. Covas has calibrated her solver by carrying out a sensitivity analysis to define model parameters. Afterwards, calibrations were carried out for several transient tests considering laminar and smooth turbulent flow: $Q_0 = 0.056\ \mathrm{l.s^{-1}}$ to $1.98\ \mathrm{l.s^{-1}}$. The wave-speed is $C_f = 385\ \mathrm{m.s^{-1}}$ to $425\ \mathrm{m.s^{-1}}$ according to the estimated values based on maximum pressure peaks and travelling wave times. The elastic wave speed was fixed at $395\ \mathrm{m.s^{-1}}$ (*i.e.* $E_0 = 1.4\ \mathrm{GPa}$ and $J_0 = 0.70\ \mathrm{GPa^{-1}}$) as a compromise between accuracy and computational time, and the relaxation times are equal to 0.05, 0.5, 1.5, 5 and 10 s. The best-fitted parameters, and the respective creep curves and calculated piezometric heads were detailed. It was concluded that the minimum number of three elements is essential to achieve an acceptable level of accuracy, and greater numbers (4 ou 5) do not improve the accuracy of the results, providing only a different combination of parameters J_k and requiring more computational time. In addition, the relaxation times $\tau_1 = 0.05\ \mathrm{s}$ and $\tau_2 = 0.5\ \mathrm{s}$ must be included in the K-V model to achieve a good accuracy of the model, and any of the others higher than the previous ones (*e.g.* $\tau_3 = 1.5\ \mathrm{s}$, 5 s or 10 s) can be used. According to the tests of Stoinov and Covas, it has been assumed that the higher the relaxation times are, the closer calibrated creeps are to the experimental function. The analysis of sample size ΔT was carried out to determine the best relaxation time τ_3, and it was concluded that creep functions for $\tau_3 = 10$ s are closer to the experimental curve than that for $\tau_3 = 1.5$ s and $\tau_3 = 5$ s.

Table **6.1** displays two series of viscoelastic parameters provided in [1] corresponding to two different cases used in this section to validate the numerical computation: the steady state laminar flow case $Q_0 = 0.056$ l.s^{-1} (Re $= 1400$) and the steady state turbulent flow case $Q_0 = 1.008$ l.s^{-1} (Re $= 25200$).

Table 6.1. Creep compliance functions for two cases of steady state flowrate.

Steady State Flowrate Q_0 (l/s)	Sample SizΔ ΔT (s)	Average Least Square Error (m²)	Creep Function J_k (10^{-9} Pa^{-1})		
			$\tau_1 = 0.5$ s	$\tau_3 = 1.5$ s	$\tau_3 = 10$ s
0.054	10	0.0014	0.0935	0.13	0.836
1.008	10	0.0472	0.104	0.124	0.41

The friction coupling is considered through steady friction (SF) terms expressed by the Darcy-Weisbach formula. The friction coefficient f can be obtained by the Blasius formula for turbulent flow and the Hagen-Poiseuille formula for laminar flow. Covas provided useful relationship between pipe friction and Reynolds number (Fig. **6.1**) [1]. According to this result, the friction coefficients used in this case study will be almost equal to 0.040 for the laminar flow and 0.025 for the smooth turbulent flow.

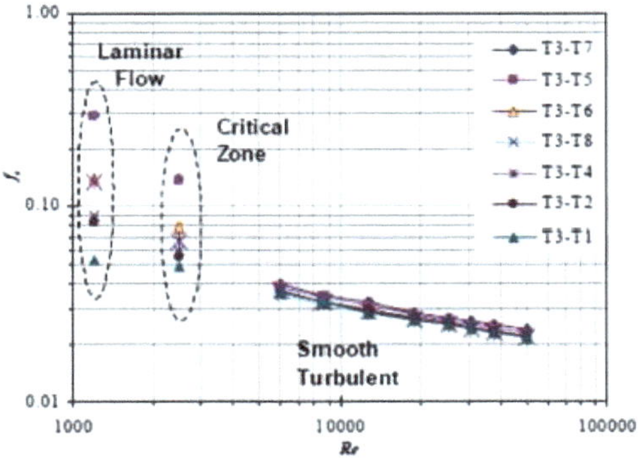

Fig. (6.1). Relationship between pipe friction and Reynolds number [1].

Case 2: Cavitation in HDPE pipeline

The experiment of Carriço, described in 5.5 of the first chapter is considered in this subsection. In fact, two tests were carried out: non cavitating flow test and cavitating flow test. The latter will be considered in this study. The generalized Kelvin-Voight model was considered to describe the creep compliance function *J*(*t*) [68]. This model is usually represented by the instantaneous elastic creep J_0 and the retarded coefficients J_k and τ_k of each element. For the cavitating flow test, the authors estimated the elastic wave-speed as $250\ \text{m.s}^{-1}$ by taking $E_0 = 1.42\ \text{GPa}$, *i.e.* $J_0 = 0.70\ \text{GPa}^{-1}$. The optimum number of Kelvin-Voight element was obtained by using three elements. The retarded times are $\tau_1 = 0.10\ \text{s}$, $\tau_2 = 0.50\ \text{s}$ and $\tau_3 = 3\ \text{s}$, and the retarded creep compliances are $J_1 = 0.60\ \text{GPa}^{-1}$, $J_2 = 0.35\ \text{GPa}^{-1}$ and $J_3 = 0.50\ \text{GPa}^{-1}$. The authors assumed unsteady friction losses to be described by the calibrated creep function. Transient was caused by the fast closure of the upstream end ball valve. In the next simulation of the cavitating flow in PE pipe, the piezometric heads at location 1 (upstream end of the pipeline and immediately downstream the ball valve) and at location 5 (middle of the pipe) will be considered.

The non-cavitating flow

Pressure analysis

The experiment of Stoinov and Covas is used to validate numerical calculations obtained by the classical VEM. The calculations are performed at three locations at the PE pipeline: transducer T1 at 270 m, transducer T5 at 116.42 m and transducer T3 at 0 m (upstream end).

The reservoir is assumed to be of time-dependent elevation. Some calibrations have been tested into the algorithm to determine the circular frequency ω and the amplitude of the wave ΔH. Figs (**6.3**) and (**6.5**) show these sin-approximations at transducer T3 for respectively laminar and turbulent flow regimes. The effect of these functions can be observed in all calculated responses where the accuracy of the results depends on them especially at the latest spikes.

Another parameter affecting pressure magnitudes is the mechanical behaviour of the pipe ends. The simulation shows that the pressure magnitudes in case of rigid ends are closer to the experimental result than those of the viscoelastic end case, whether the steady state flow is laminar (Fig. **6.2**), or turbulent Figs. (**6.4**) and (**6.6**). In fact, this discrepancy is realistic because the pipe ends are not PE made, and retarded circumferential strain at these locations should be omitted from boundary condition calculations. The effect of the retarded circumferential strain can be adopted in the model as supplementary boundary condition proposed in this work. Even with rigid end condition, slight discrepancies are shown in magnitudes. In 2016, Guidara attributed them to the negligence of UF in his thesis work. However, SF has no influence on the numerical result. Regarding dispersion, despite the good agreement exhibited by the classical VEM, a slight timing delay appears due to wave-speed error (the exact wave-speed is slightly higher than 395 $m.s^{-1}$).

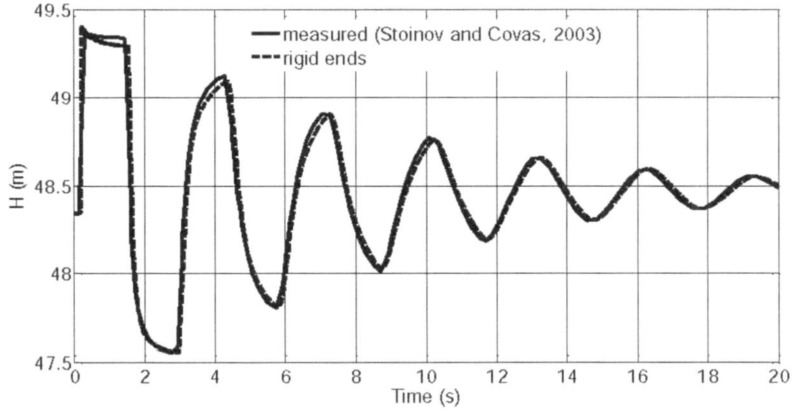

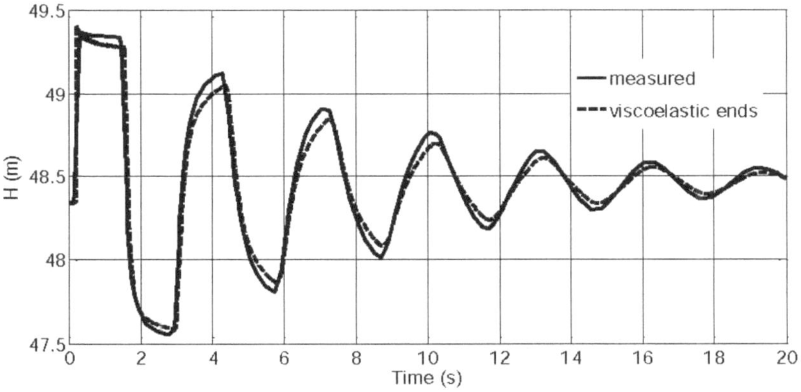

Fig. (6.2). Piezometric head at location T1 obtained by the classical VEM for Q_0 = 0.056 l/s.

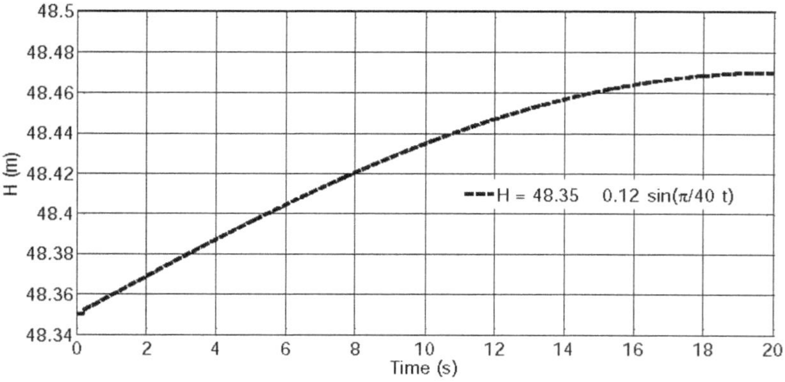

Fig. (6.3). Piezometric head at location T3 for $Q_0 = 0.056$ l/s.

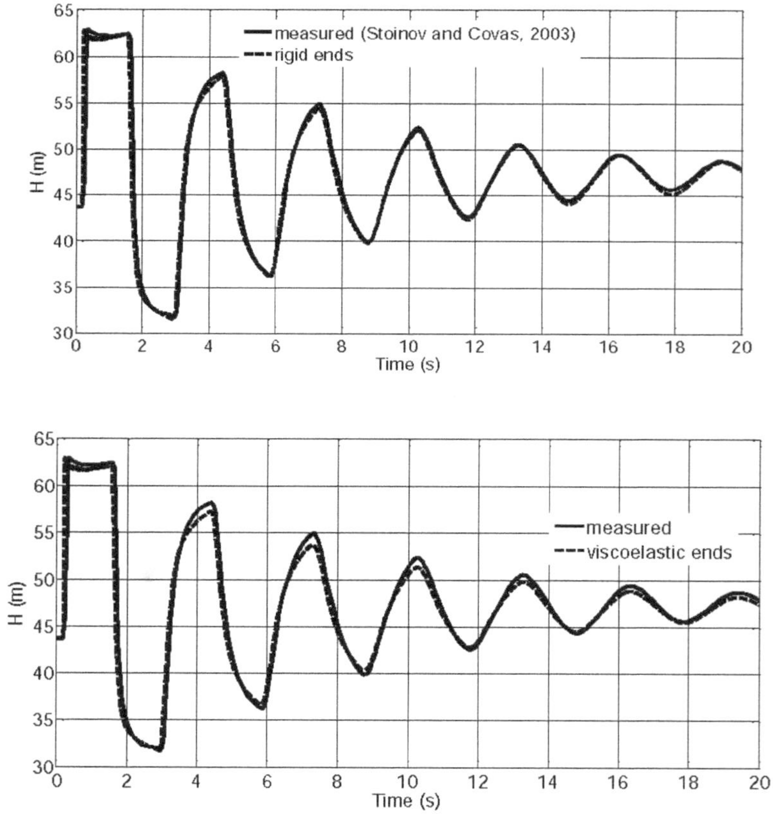

Fig. (6.4). Piezometric head at location T1 obtained by the classical VEM for $Q_0 = 1.008$ l/s.

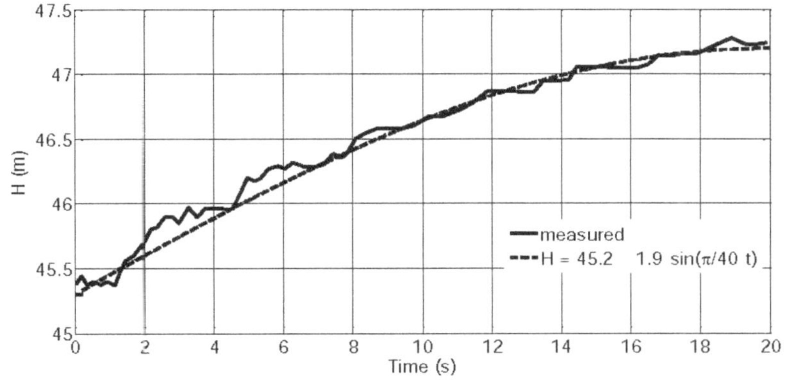

Fig. (6.5). Piezometric head at location T3 for Q_0 = 1.008 l/s.

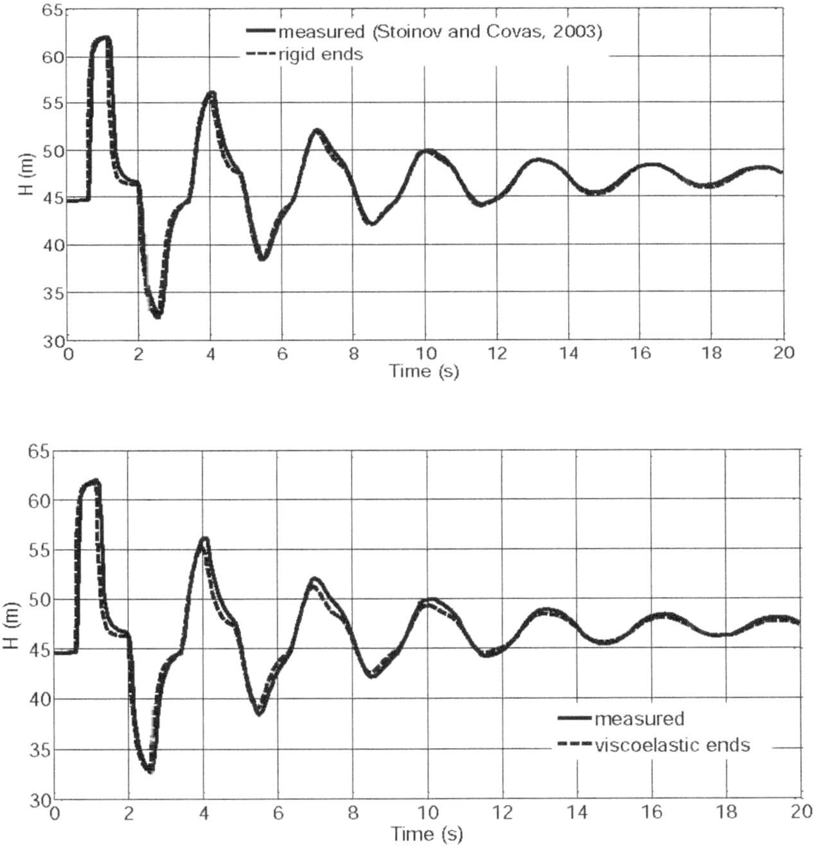

Fig. (6.6). Piezometric head at location T5 obtained by the classical VEM for Q_0 = 1.008 l/s.

Strain Analysis

Figs. (**6.7** and **6.8**) show the circumferential strain histories at locations T1 and T5, respectively. The same description given below for piezometric head histories is still valid for the circumferential strains, namely the effect of the end mechanical behaviour. Obviously, the rigid end assumption leads to more accuracy regarding strain magnitudes. As mentioned above, this latter assumption seems more realistic than the viscoelastic end assumption. However, a shape discrepancy is exhibited at the first pressure rise for the two cases of (Fig. **6.7**), which is likely due to the UF effect that was neglected in the classical VEM. Since the pipe is rigidly fixed against axial vibration, the numerical results provided by the classical VEM are in good agreement with the experimental result.

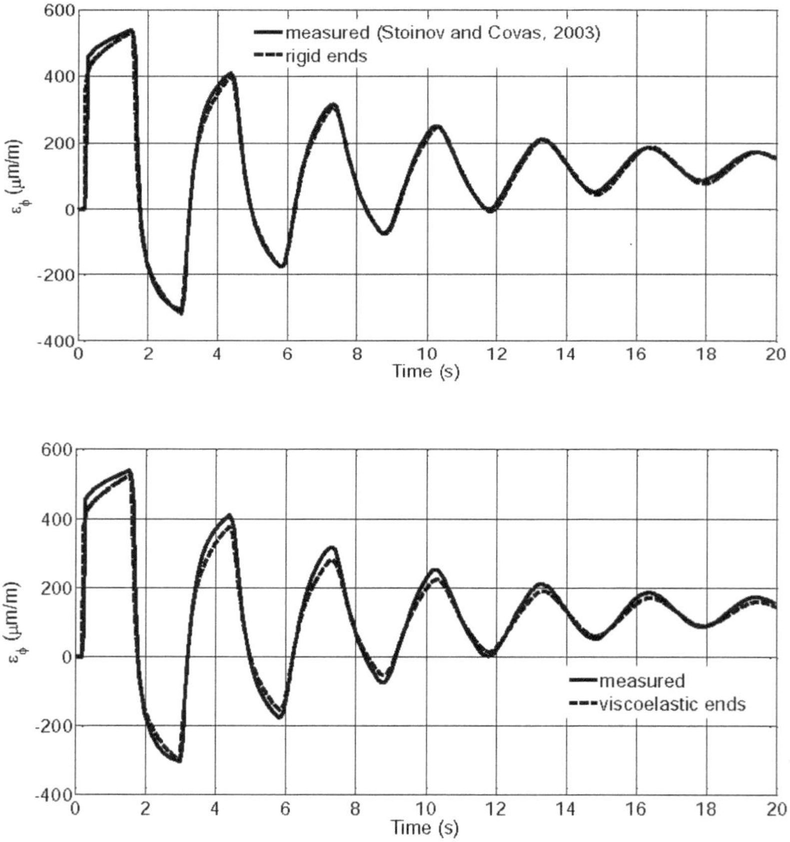

Fig. (6.7). Circ. strain at location T1 obtained by the classical VEM for Q_0 = 1.008 l/s.

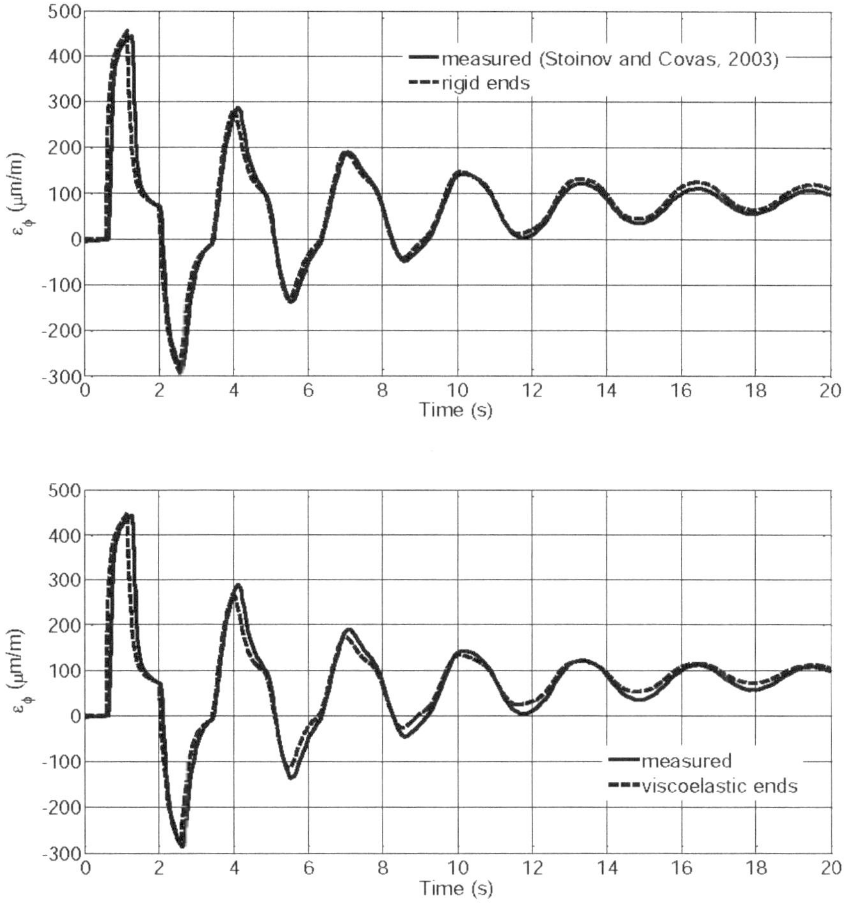

Fig. (6.8). Circ. strain at location T5 obtained by the classical VEM for Q_0 = 1.008 l/s.

The Cavitating Flow

The cavitating flow case is considered as mentioned in 2.2, where the experiment of Carriço was carried out for a rigidly restrained HDPE pipe. There was no free axial movement of the pipe so that junction coupling is ignored in this test. First, the simulation of transient is obtained by use of earlier elastic models: the classical DVCM, the classical DGCM, the coupled DVCM and the coupled DGCM. The objective is to exhibit the failure of this models in prediction of column separation in VE pipes. Then the viscoelastic models are validated against experimental records.

Deficiency of the Elastic Models

Figs. (**6.9**) to (**6.11**) display the deficiency of the elastic models in prediction of column separation in viscoelastic pipelines. A small number of reach $N = 16$ is used for the simulation. Fig. (**6.9**) shows the failure of the elastic DVCM to simulate the damping effect caused by viscoelasticity. Although it does not exhibit great lag, the magnitude error is considerably large. The elastic DGCM is not better than the elastic DVCM except the slight magnitude attenuation that it provides (Fig. **6.10**).

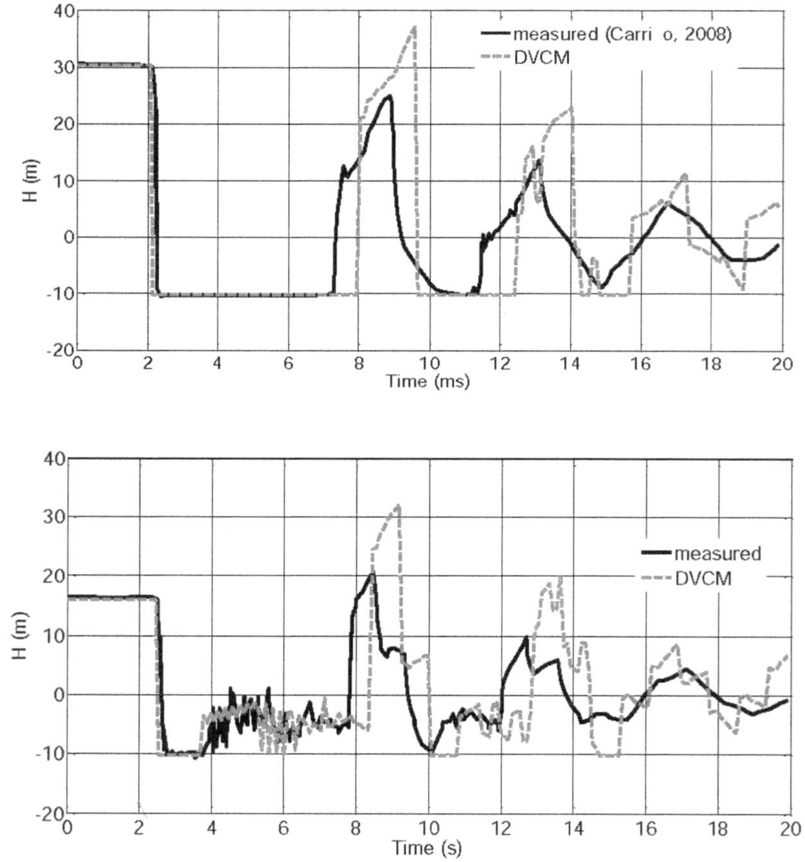

Fig. (6.9). Piezometric heads calculated by the elastic DVCM. Top: at location T1; bottom: at location T5.

The calculation obtained with the coupled elastic DVCM is plotted in Fig. (**6.11**) where the upstream valve is rigidly clamped to the reservoir (no junction coupling).

The characteristic direction ratio is equal to 5.51, so that the rational number 11/2 is used to calculate variables in the WSA scheme ($a = 2$ and $b = 11$).

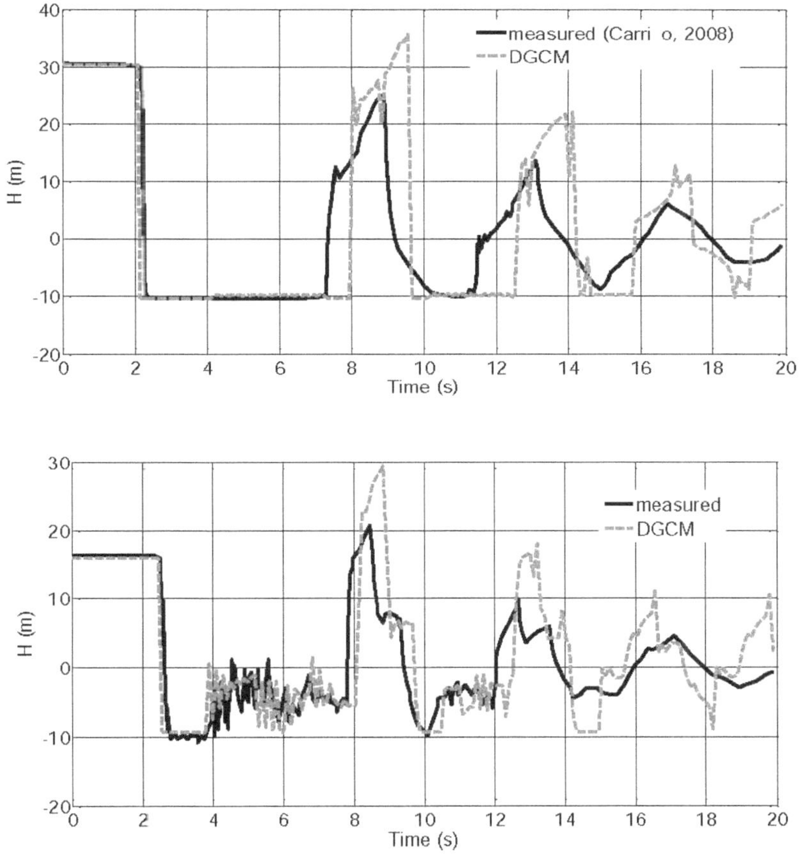

Fig. (6.10). Piezometric heads calculated by the elastic DGCM. Top: at location T1; bottom: at location T5.

The earlier column separation models used for elastic pipes are tested in the present case of cavitation in viscoelastic pipelines. Some physical properties of the HDP are not mentioned in the experiment of Carriço, such as the mass density and the Poisson coefficient. The two values attributed and used in calculation are $\rho_p = 960$ kg.m^{-3} and $\nu = 0.46$. Also, a pipe thickness of 3.2 mm is used. The pressure head of the inlet reservoir is approximated at 1 m, whereas a constant head of 30.5 m is attributed to the upstream reservoir. The friction coefficient can be calculated either using the Moody chart or the Colebrook equation. Since the flow is turbulent and the pipe is assumed to be smooth, the friction coefficient can be calculated by the formula $f = 0.316\,\mathrm{Re}^{0.25}$, which gives $f \approx 0.018$. Unlike column separation in elastic pipelines that has been successfully predicted by this model, the simulation exhibits higher magnitudes compared to the classical model. The timing delay is also increased. The main origin of discrepancy exhibited by the elastic models is the viscoelastic behaviour of the pipe.

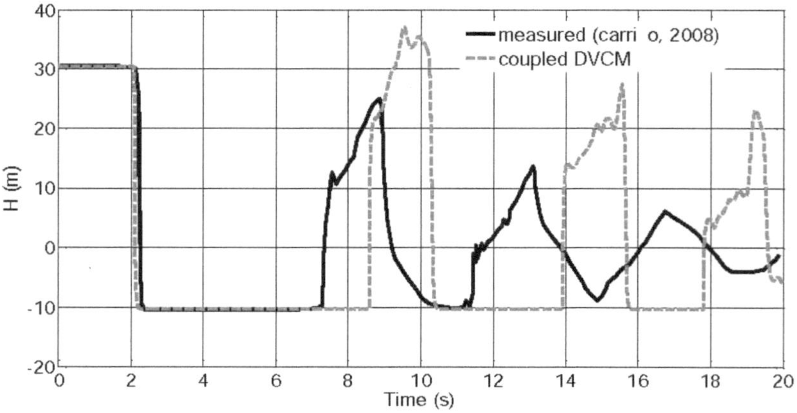

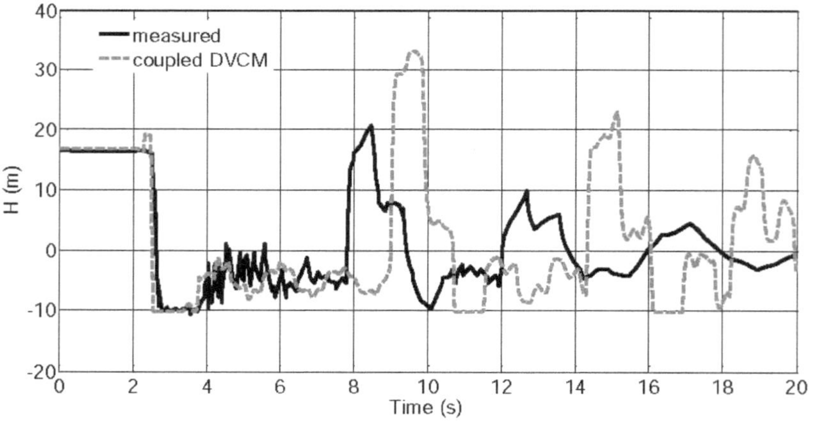

Fig. (6.11). Piezometric heads calculated by the elastic coupled DVCM. Top at location T1; bottom: at location T5.

Efficiency of the Viscoelastic Models

Unlike the disagreement exhibited by the elastic models with the experimental results, the viscoelastic models can give good results if the appropriate parameters are considered. In this study, two parameters are involved: (i) the mechanical behaviour at the pipe ends and (ii) the creep-compliance functions.

Effect of the Pipe Ends

Figs. (**6.12**) to (**6.16**) show good agreement between calculation and experiment. The mechanical behaviour at the pipe ends is considered as in section 2: rigid ends and VE ends.

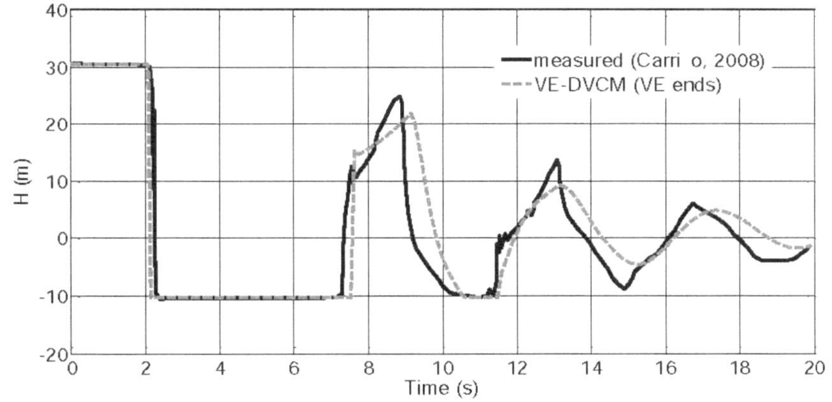

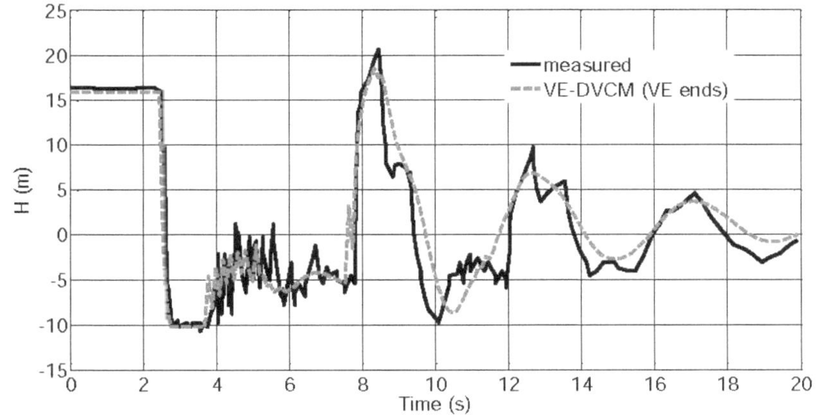

Fig. (6.12). Piezometric head calculated by the VE-DVCM with VE-end condition. Top at the upstream valve (location T1), bottom: at the midstream section (location T5).

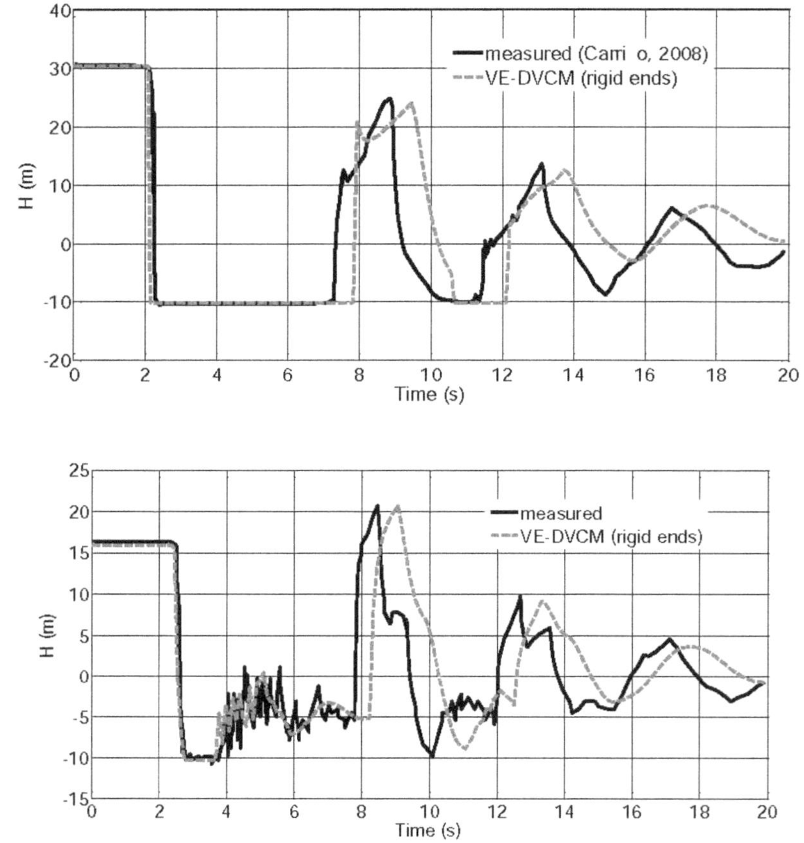

Fig. (6.13). Piezometric head calculated by the VE-DVCM with rigid-end condition. Top at the upstream valve, bottom: at the midstream section.

The simulation of the VE-DVCM Figs. (**6.12**) and (**6.13**) displays comparison between these two conditions. Although the rigid-end condition is more realistic and more accurate regarding magnitudes, the VE-end condition leads to more accuracy in timing. The same comparison established for the VE-DVCM is valid for simulation of the VE-DGCM displayed in Figs. (**6.14**) and (**6.15**). Besides, the rigid-end condition avoids the discrepancy of pressure in the cavity exhibited by the VE-end condition (Fig. **6.14**). Moreover, the lag observed for the VE-DVCM disappears by use of the VE-DGCM, which allows the preference of the rigid-end condition to the VE-end condition.

Regarding circumferential strain histories, Figs (**6.20**) to (**6.22**) show the effect of end behaviour on the strain at the middle of the pipe. The rigid-end condition considered for the VE-DVCM allows circumferential strain to reach zero at time $t = 9$ s, which corresponds to the first pressure spike of the rigid end in Fig. (**6.12**) (bottom). By use of the VE-end condition, the strain cannot exceed -2 mm/m, at time $t = 8.5$ s, because at this time, the pressure spike of the VE-end condition is lesser than the former. The contracting and expanding behaviour of the pipe is caused by the pressure variation: the pressure drop results in pipe contraction and the pipe expands because of a pressure rise.

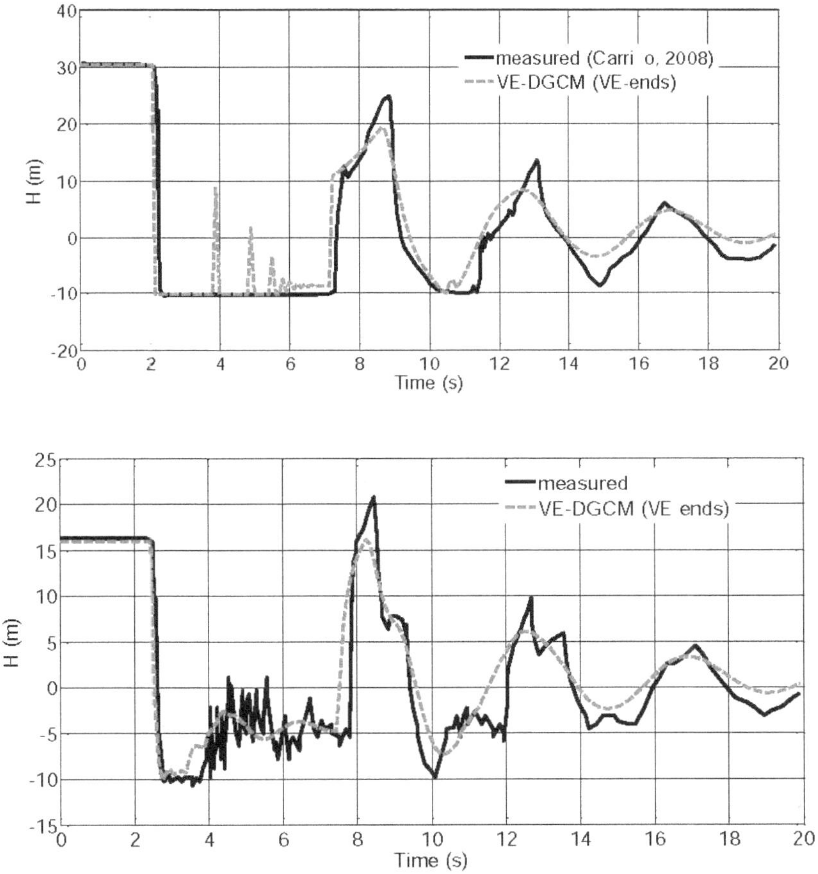

Fig. (6.14). Piezometric head calculated by the VE-DGCM with VE-end condition. Top at the upstream valve, bottom: at the midstream section.

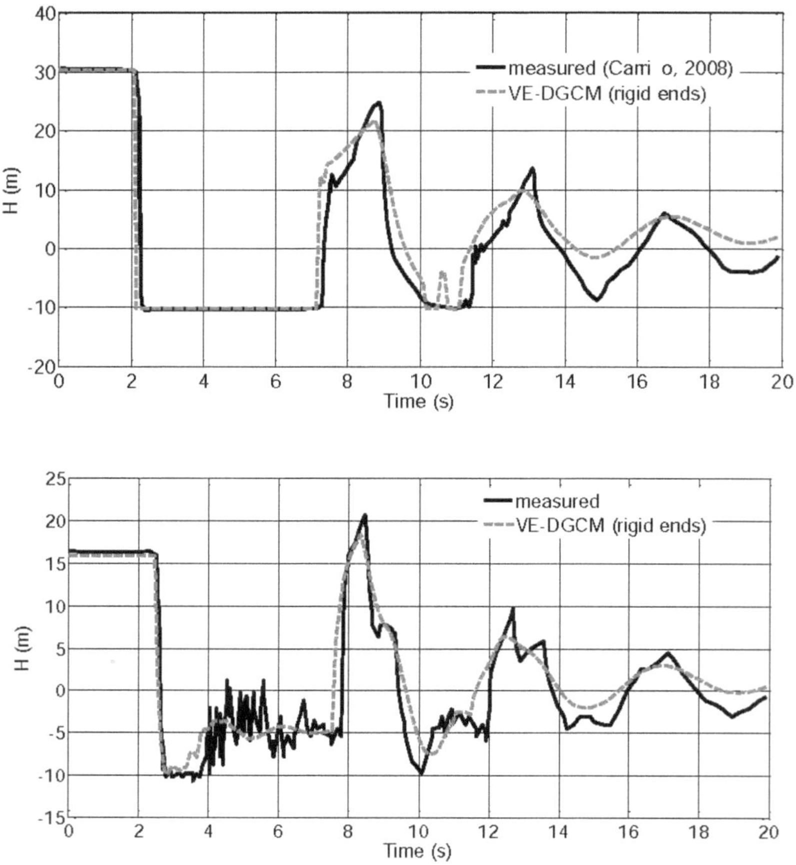

Fig. (6.15). Piezometric head calculated by the VE-DGCM with rigid-end condition. Top at the upstream valve, bottom: at the midstream section.

Fig. **(6.16)** shows the effect of gas release on the circumferential strain and compares the VE-DGCM to the VE-DVCM in case of rigid and VE-ends. The discrepancy is related to the pressure rises described previously. This can be explained by the effect of released gas in the cavity on the collapse. For the VE-DGCM, the pressure rise following the collapse is reduced which leads to less expansion of the pipe.

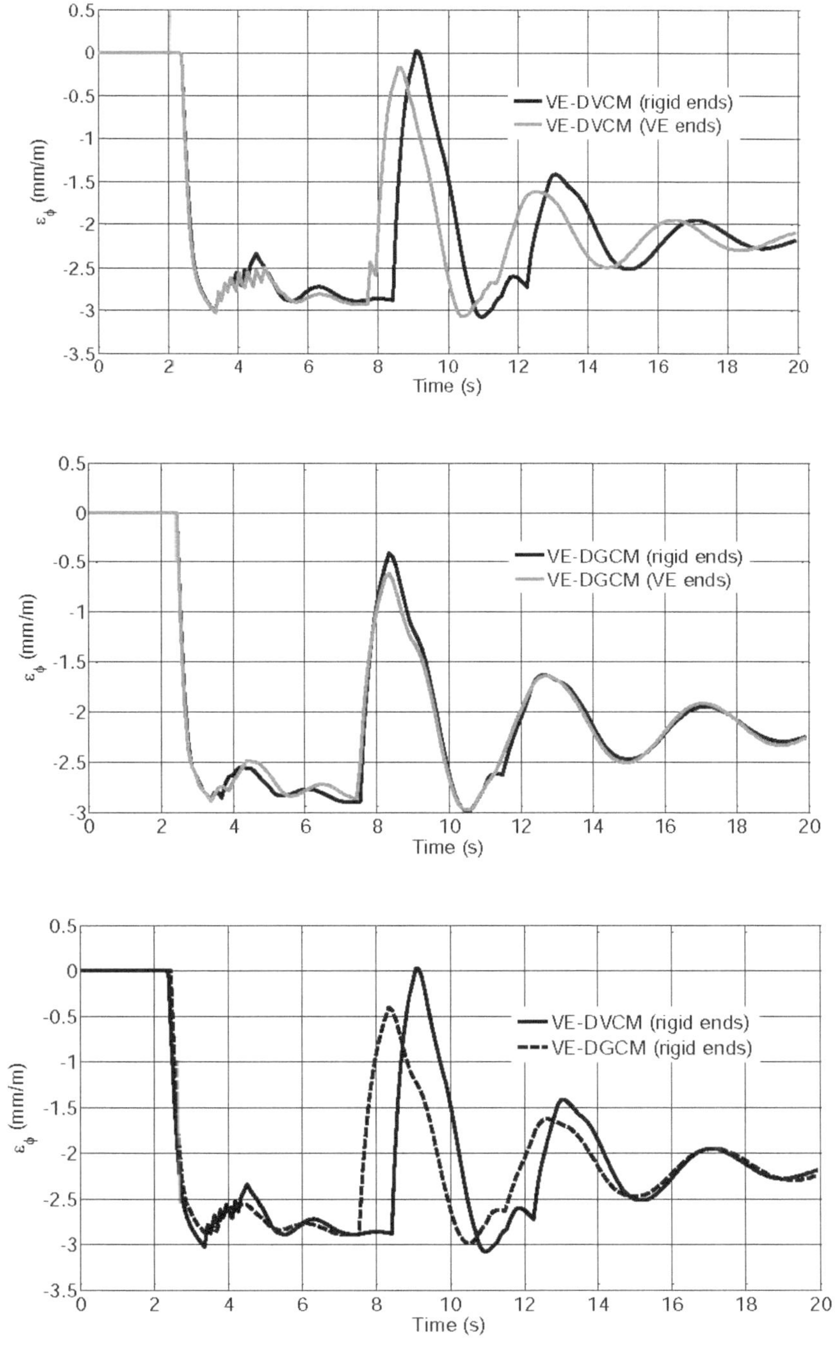

Fig. (6.16). Circumferential strain at the midstream with different models.

Effect of the Creep-Compliance Functions

In the earlier research, the creep-compliance functions are assumed to be constant during fluid transient with cavitation. However, these functions can be either constant or variable according to cavitation occurrence. In the experimental tests of Carriço, there are two sets of creep-compliance functions: one for water hammer case and the other for cavitation case. The previous calculation is performed by considering the constant creep-compliance (CCC) approach, in which the creep-compliance functions are defined in paragraph 1.2. The present subsection involves the variable creep-compliance (VCC) approach and gives analysis of its effect on the result (Fig. **6.17**).

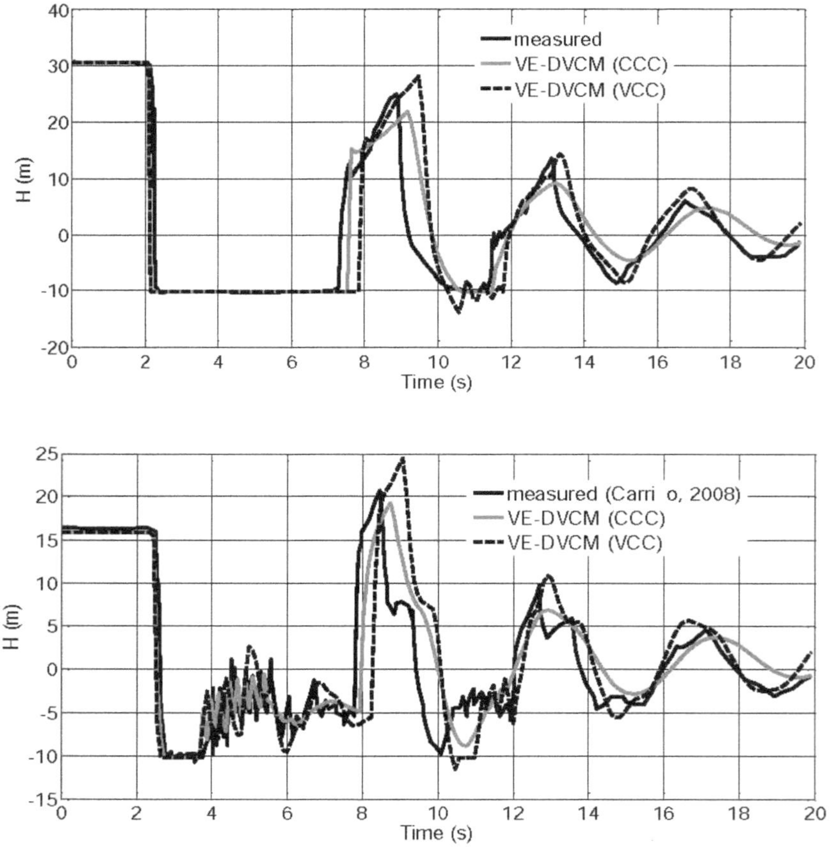

Fig. (6.17). Piezometric head calculated by the VE-DVCM with VE-end condition. Top: at the upstream valve, bottom: at the midstream section.

The retarded parameters of the cavitating case are defined in paragraph 1.2, whereas those of the non-cavitating case (water hammer case) are $\tau_1 = 0.018$ s, $\tau_2 = 0.50$ s, $\tau_3 = 3$ s, $J_1 = 0.256$ GPa^{-1}, $J_2 = 0.238$ GPa^{-1} and $J_3 = 0.290$ GPa^{-1} [68].

Figs. **(6.17)** and **(6.18)** compare the piezometric heads calculated with the VE-DVCM of the above approaches in case of the VE-end and the rigid-end conditions, where a limited number of reaches $N = 16$ is used for all computation. Regardless the end condition used, the calculation shows the preference of the VCC condition against the CCC condition, and this preference is especially exhibited in magnitudes rather than in timing lag.

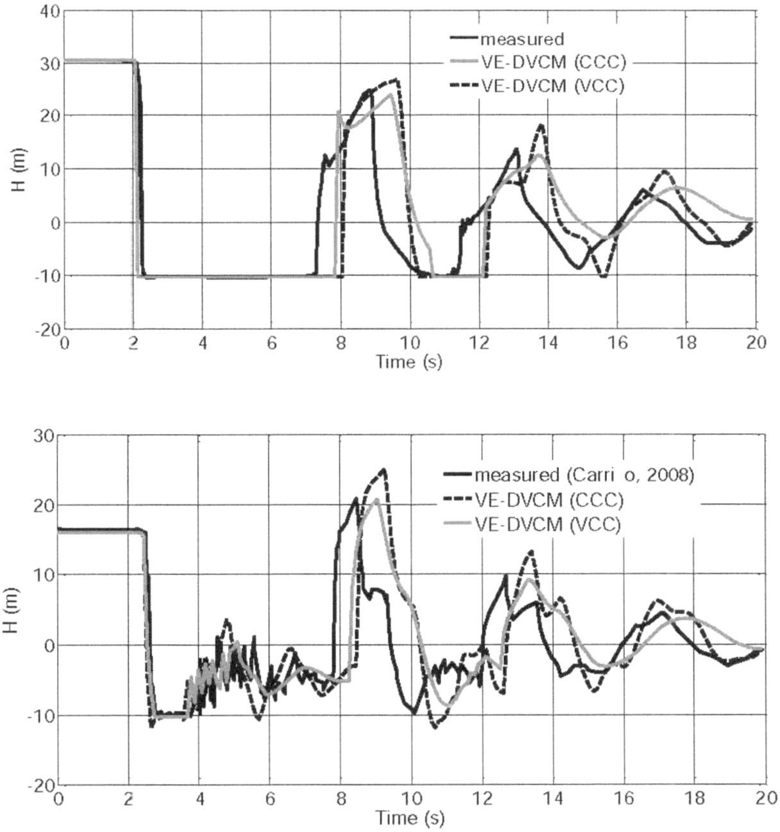

Fig. (6.18). Piezometric head calculated by the VE-DVCM with rigid-end condition. Top: at the upstream valve, bottom: at the midstream section.

Figs. (**6.19**) and (**6.20**) display the results of the VE-DGCM. The two approaches of the creep-compliance functions are compared again for both VE-end and rigid-end conditions. The VCC approach is still preferred to the CCC approach regarding magnitudes for both rigid and VE-ends. A timing error is however involved by use of The VCC with VE-ends (Fig. **6.19**). The CCC approach provides numerical solution with acceptable timing, but it does not correctly predict the wave attenuation; the damping effect is higher than that of the VE-DVCM (Fig. **6.17**). Regarding shape, it the solution provided by the VCC approach with VE-end condition represents the best method to simulate the pressure history at the valve and at the midstream section of the HDPE pipe.

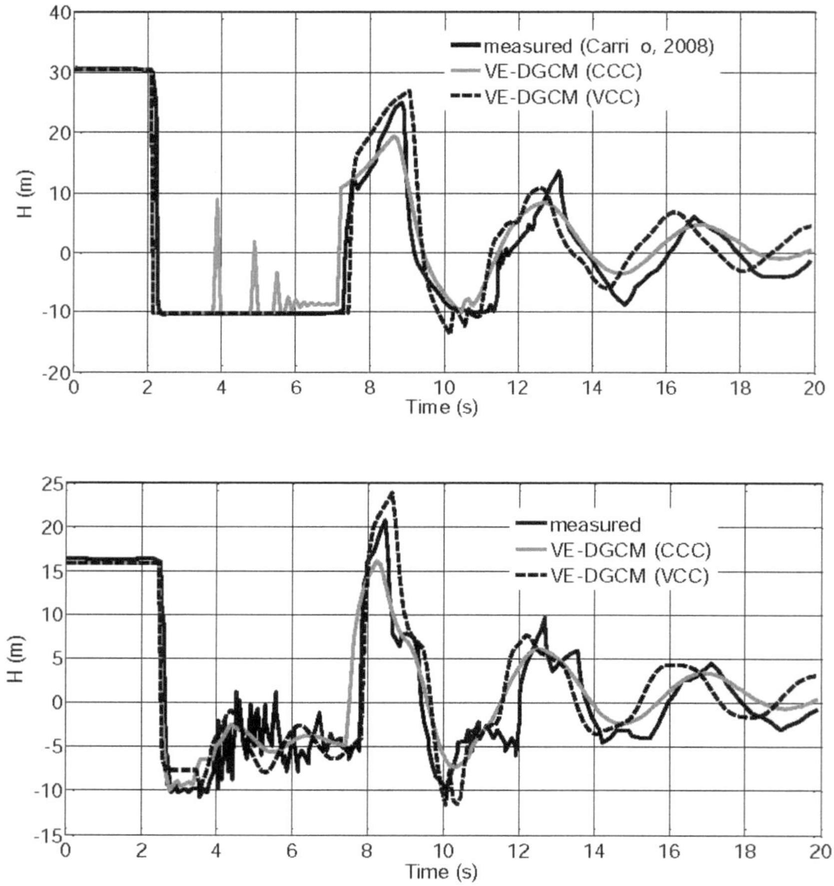

Fig. (6.19). Piezometric head calculated by the VE-DGCM with VE-end condition. Top: at the upstream valve, bottom: at the midstream section.

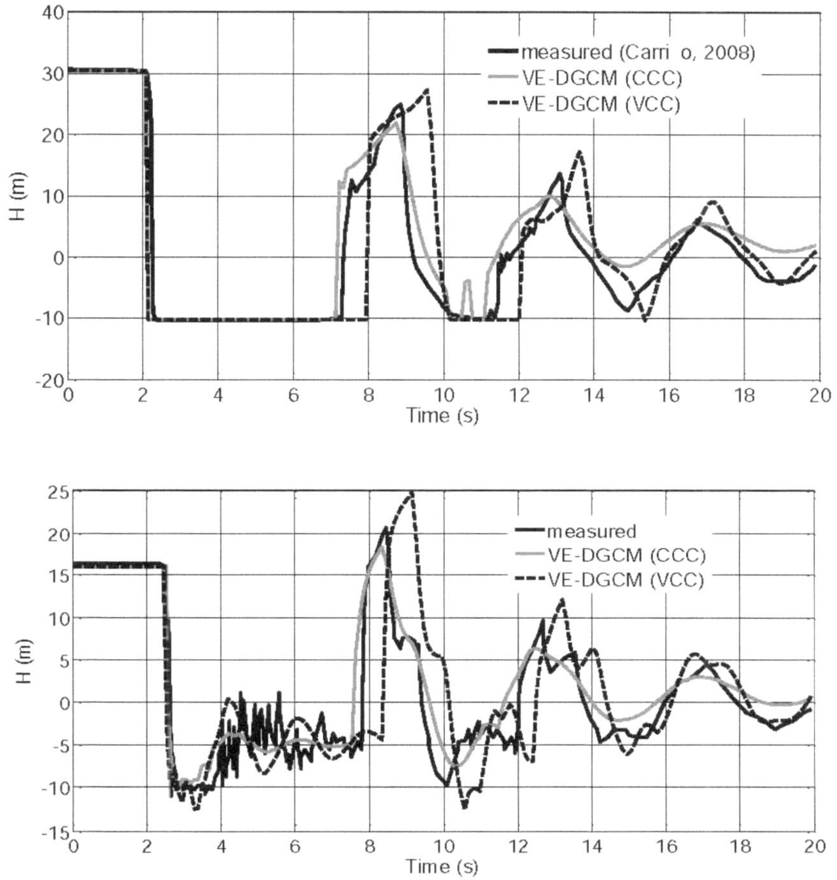

Fig. (6.20). Piezometric head calculated by the VE-DGCM with rigid-end condition. Top: at the upstream valve, bottom: at the midstream section.

CONCLUSION

Transient in viscoelastic pipelines has been studied in the present chapter, where two cases have been discussed: the non-cavitating flow and the cavitating flow. The case studies used to validate the result has not consider the axial movement of the pipe, so that junction coupling has not been considered. The simulation considers two boundary conditions, which affect the result: the rigid-end condition and the VE-end condition. In the experiment of Stoinov and Covas, transient is caused by the fast closure of a downstream valve, while an upstream valve causes cavitation in the experiment of Carriço. In case of water hammer without cavitation (case 1), the classical VEM is sufficient to predict column separation in HDPE pipeline. This is because junction coupling has been avoided and Poisson coupling modelling has no significant effect in case of restrained pipeline. The simulation shows that the result obtained by the classical VEM is in good agreement with the experimental results. In case of cavitating flow (case 2), the calculation shows deficiency of the elastic models and efficiency of the viscoelastic models of cavitation. The coupled elastic DVCM is not better than the classical elastic models. This can be due to the pipe anchoring and absence of axial vibration. The viscoelastic effect (damping and dispersion) is more relevant than the Poisson coupling effect. The VE-DVCM as well as the VE-DGCM have been successfully used to predict column separation in HDPE pipes. Their solutions can be improved by use of the variable creep-compliance (VCC) method. The simulation shows that the VE-DGCM simulates better column separation especially when the rigid-end condition is considered.

REVIEW AND CONCLUSIONS

In this work, fluid-structure interaction (FSI) modelling and numerical simulation of transient in pipelines have been carried out. The emphasis was made on the development of column separation models taking into account several parameters such as friction, Poisson coupling, junction coupling (axial movement of the pipe) and viscoelasticity. The derived equations are valid for long wavelength and hence low frequency. All mathematical and numerical modelling are based on the standard reservoir-pipe-valve system, where transient is caused by the fast closing of the downstream valve, except in case of cavitation in viscoelastic pipelines for which transient is caused by the quick closure of the upstream valve.

An extensive review of literature has been carried out. It has given description of physical phenomenon and mathematical and numerical modelling of hydraulic transients in pipelines. Several experiments on water hammer and cavitation in elastic and viscoelastic pipelines have been investigated. The proposed models of water hammer in pipelines have been developed based on the continuity equation and the momentum equation for both the fluid and the pipe. The modelling takes account of elasticity, unsteady friction, and viscoelasticity. Thus, three coupled models have been detailed allowing calculation of the pressure and the axial velocity of the fluid and the axial stress and the axial velocity of the pipe. These coupled models are in fact FSI models that take into account three types of dynamic coupling: friction coupling, Poisson coupling and junction coupling. The four-equation model (4EM) and the four-equation friction model (4EFM) describe water hammer in elastic pipes with free axial vibration. The numerical modelling of cavitation has been based on the water hammer models and the classical column separation models, namely the classical discrete vapour cavity model (DVCM) and the classical discrete gas cavity model DGCM. Both vaporous and gaseous cavitation have been studied by considering only column separation at specific locations along the pipeline. Distributed cavitation models have not been studied in this work.

Regarding the numerical procedure, the MOC has been used to solve the water hammer models in both elastic and viscoelastic models. The WSA scheme has been preferred to both time-line interpolation (TLI) and space-line interpolation (SLI) schemes. The MOC development based on the WSA scheme had been first applied

Abdelaziz Ghodhbani, Ezzeddine Haj Taïeb, Mohsen Akrout & Sami Elaoud

to the water hammer elastic models (the 4EM and the 4EFM) and then extended to cavitation.

The cavitation study has only considered column separation phenomenon, which is dominant in hydraulic plants. The cavitation modelling has been obtained by referring to the water hammer models. The compatibility equations of the water hammer models have been used to develop the coupled DVCM and the coupled DGCM for elastic pipelines. Therefore, the same approach has been extended to the viscoelastic pipelines. The MOC development has been also used to solve the coupled viscoelastic model referred to as the 4EVEM. This latter has needed further development because of the existence of the retarded strain in the governing equations. As proposed for the elastic pipelines, the compatibility equations of this model leaded to the coupled viscoelastic DVCM, referred to as coupled VE-DVCM. Moreover, gas release phenomenon was taken into account by providing the coupled DGCM which was only applied for elastic pipes.

The valve anchoring conditions were also considered in the coupled numerical models. For each model, the boundary condition at the valve was considered in two different modes. The first one corresponds to the fixed valve which ignores junction coupling. The second which is more complicated in calculation corresponds to the freely moving valve. This latter supposes that the valve can move axially so that the material of the support (whether rigid or viscoelastic) matters.

The developed models have been implemented in software package using computer codes developed by the present authors. The implementation of the numerical models into the software package using matrix transformation is detailed in appendix D where only the downstream-valve system is considered. The simulation and the manual calibration lead to the following conclusions:

i) The instantaneous closure valve is a successful way to cause transient; the discrepancy that it exhibits is negligible compared to the non-instantaneous closure valve, which causes nonlinearity and needs numerical method to solve.
ii) The WSA scheme is more reliable and more suitable than the interpolation schemes although some manual calibration are needed during calculation, and in addition, the density adjustment are not necessary, and they do not influence the result.
iii) The classical models of water hammer are as accurate as junction coupling is reduced. The rigid the pipe anchor is, the efficient the classical model is.

iv) The weakness of the classical column separation models (the DVCM and the DGCM) is their dependence to the mesh refinement and their non-convergence. Also, these models cannot predict structural responses.

v) The FSI models such as the four-equation model (4EM), the four-equation friction model (4EFM), the coupled DVCM and the coupled DGCM are preferred to the classical models regarding accuracy, convergence and computational efficiency. The 4EM 4EFM are preferred especially in case of axially vibrating pipes with quasi-rigid or viscoelastic supports.

vi) The main advantage of the FSI models is their ability to predict structural responses such as axial displacement, axial strain, and axial stress. The structural responses can be assumed to be accurate once the pressure history is validated against experimental results.

vii) In case of cavitation in elastic pipelines, the FSI models provide more accuracy even when the pipe is rigidly anchored. The coupled DVCM and the coupled DGCM proposed in this work have been validated in case of hydraulic system with fixed valve, but the proposed approach is also valid for freely moving pipes.

viii) The simulation shows that the support material has an important impact on fluid and structural responses during fluid transients in elastic pipelines, but the numerical simulation has not been validated with experimental results.

ix) Unsteady friction (UF) is more suitable because it yields more damping and more accuracy in elastic pipes. In viscoelastic pipes, its damping effect is negligible compared to the viscoelastic damping, and the calculation can be limited to the steady-state friction.

x) Gas release leads to more accuracy when considered whatever the model and whatever the pipe material (classical DGCM and coupled DGCM for elastic pipes and VE-DGCM for VE pipes).

xi) The classical viscoelastic models (VEM) for water hammer and cavitation can provide accurate results in case of anchored pipeline, and on would not need coupled models for this case, such as the 4EVEM and the coupled VE-DVCM, which can be more suitable in case of vibrating structures.

xii) Some structural responses such as the circumferential strain and the hoop stress can be accurately predicted with the classical VEM in case of rigidly anchored pipes.

xiii) Cavitation prediction in VE pipelines can be improved if the variable creep-compliance (VCC) approach is used instead of the constant creep-compliance (CCC) approach. The main advantage of the former consists in pressure magnitudes.

The modelling and the numerical simulation proposed for water hammer and cavitation in pipelines are analysed and detailed taking the most hydraulic and mechanical cases that can be observed in practice. Nonetheless, the study has not considered turbomachines (pumps and turbines) which represent important components in hydraulic plants and pipeline systems. The developed models have been validated against experimental results in case of valve-closure induced transients. Pump and turbine stopping, and failure and other events can be mathematically modelled and incorporated into the above models to predict fluid transients.

Appendices

Appendix A: Generalized Hook's Law

The generalized Hook's law is attributed to Robert Hooke who established in 1678 that the deformation of a structure is proportional to the applied forces. This law is valid for solid material with linear elastic behaviour. For one-dimensional configuration, the Hook's law is written by the linear relationship $\sigma = E\varepsilon$, in which avec σ is the strength stress, ε is the strain and E is the Young's modulus of elasticity.

In the three-dimensional system, σ and ε are no longer scalars but they are 2nd order tensors. The generalized Hook's law is defined in the cylindrical coordinate system as:

$$\begin{bmatrix} \sigma_r \\ \sigma_\varphi \\ \sigma_z \\ \tau_{r\varphi} \\ \tau_{\varphi z} \\ \tau_{zr} \end{bmatrix} = \begin{bmatrix} C_{11} & C_{12} & . & . & . & C_{16} \\ C_{21} & . & . & . & . & . \\ . & . & . & . & . & . \\ . & . & . & . & . & . \\ . & . & . & . & . & . \\ C_{61} & . & . & . & . & C_{66} \end{bmatrix} \begin{bmatrix} \varepsilon_r \\ \varepsilon_\varphi \\ \varepsilon_z \\ 2\varepsilon_{r\varphi} \\ 2\varepsilon_{\varphi z} \\ 2\varepsilon_{zr} \end{bmatrix} \quad \textbf{(A.1)}$$

which can be expressed as:

$$\sigma_{ij} = C_{ijkl}\varepsilon_{kl} \quad \textbf{(A.2)}$$

with **C** is the 4th order tensor of stiffness. Since $\boldsymbol{\sigma_{ij}}$ and $\boldsymbol{\varepsilon_{kl}}$ are symmetric, the components $\mathbf{C_{ijkl}}$ are reduced to 36. For homogeneous isotropic materials, the generalized Hook's law is:

Abdelaziz Ghodhbani, Ezzeddine Haj Taïeb, Mohsen Akrout & Sami Elaoud

$$\sigma_{ij} = \lambda \varepsilon_{kk} \delta_{ij} + 2\mu \varepsilon_{ij} \quad \textbf{(A.3)}$$

with λ denotes the Lamé coefficient, μ is the stiffness modulus, ε_{kk} is the trace of the matrix $\boldsymbol{\varepsilon}_{kl}$ given by $\varepsilon_{kk} = \varepsilon_r + \varepsilon_\varphi + \varepsilon_z$, and δ_{ij} is the Kronecker's index. The development of Eq. (A.3) leads to:

$$\sigma_{ij} = \lambda \varepsilon_{kk} \delta_{ij} + 2\mu \varepsilon_{ij}$$

$$\sigma_r = \lambda \varepsilon_{kk} + 2\mu \varepsilon_r$$

$$\sigma_\varphi = \lambda \varepsilon_{kk} + 2\mu \varepsilon_\varphi$$

$$\sigma_z = \lambda \varepsilon_{kk} + 2\mu \varepsilon_z \quad \textbf{(A.4)}$$

$$\tau_{r\varphi} = 2\mu \varepsilon_{r\varphi}$$

$$\tau_{\varphi z} = 2\mu \varepsilon_{\varphi z}$$

$$\tau_{zr} = 2\mu \varepsilon_{zr}$$

It follows:

$$\sigma_{kk} = (3\lambda + 2\mu)\varepsilon_{kk} \quad \textbf{(A.5)}$$

The incorporation of Eq. (A.5) into Eq. (A.3) allows:

$$\varepsilon_{ij} = \frac{1}{2\mu}\left(\sigma_{ij} - \frac{\lambda}{3\lambda + 2\mu}\sigma_{kk}\delta_{ij}\right) \quad \textbf{(A.6)}$$

which is usually written as:

$$\varepsilon_{ij} = \frac{1+\nu}{E}\sigma_{ij} - \frac{\nu}{E}\sigma_{kk}\delta_{ij} \quad \textbf{(A.7)}$$

With E is the Young's modulus of elasticity given by $E = \mu(3\lambda + 2\mu)/(\lambda + \mu)$ and ν is the Poisson's coefficient defined by $\nu = \lambda/\left[2(\lambda + \mu)\right]$. In the cylindrical coordinate system, Eq. (A.7) is:

$$\varepsilon_r = \frac{\partial u_r}{\partial r} = \frac{1}{E}\left[\sigma_r - \nu\left(\sigma_\varphi + \sigma_z\right)\right] \qquad \textbf{(A.8)}$$

$$\varepsilon_\varphi = \frac{u_r}{r} = \frac{1}{E}\left[\sigma_\varphi - \nu\left(\sigma_r + \sigma_z\right)\right] \qquad \textbf{(A.9)}$$

$$\varepsilon_z = \frac{\partial u_z}{\partial z} = \frac{1}{E}\left[\sigma_z - \nu\left(\sigma_r + \sigma_\varphi\right)\right] \qquad \textbf{(A.10)}$$

$$\varepsilon_{r\varphi} = \frac{1+\nu}{E}\tau_{r\varphi} \qquad \textbf{(A.11)}$$

$$\varepsilon_{\varphi z} = \frac{1+\nu}{E}\tau_{\varphi z} \qquad \textbf{(A.12)}$$

$$\varepsilon_{zr} = \frac{1+\nu}{E}\tau_{zr} \qquad \textbf{(A.13)}$$

The above equations lead to:

$$\sigma_r = \frac{E}{(1+\nu)(1-2\nu)}\left[(1-\nu)\varepsilon_r + \nu\left(\varepsilon_\varphi + \varepsilon_z\right)\right] \qquad \textbf{(A.14)}$$

$$\sigma_\varphi = \frac{E}{(1+\nu)(1-2\nu)}\left[(1-\nu)\varepsilon_\varphi + \nu\left(\varepsilon_r + \varepsilon_z\right)\right] \qquad \textbf{(A.15)}$$

$$\sigma_z = \frac{E}{(1+\nu)(1-2\nu)}\left[(1-\nu)\varepsilon_z + \nu\left(\varepsilon_r + \varepsilon_\varphi\right)\right] \qquad \textbf{(A.16)}$$

and subsequently,

$$\sigma_r = E\varepsilon_r + \nu\left(\sigma_\varphi + \sigma_z\right) \qquad \textbf{(A.17)}$$

$$\sigma_\varphi = \frac{E}{1-\nu^2}\left(\varepsilon_\varphi + \nu\varepsilon_z\right) + \frac{\nu}{1-\nu}\sigma_r \tag{A.18}$$

$$\sigma_z = \frac{E}{1-\nu^2}\left(\varepsilon_z + \nu\varepsilon_\varphi\right) + \frac{\nu}{1-\nu}\sigma_r \tag{A.19}$$

Appendix B: Stress Distribution in a Pressurized Ring

In this appendix, a circular section pipe of inner radius R and wall thickness e is considered. A constant pressure p is maintained inside the pipe while the pressure p_{out} (like the barometric pressure) is assumed to exist outside. In order to express radial and hoop stress in the pipe, a thin ring is considered (Fig. **B.1**). The bi-harmonic Airy's function Φ can be used in polar coordinate system. By neglecting the inertial effect of both the fluid and the pipe, the Airy's function satisfies:

$$\Delta^4 \Phi = 0 \tag{B.1}$$

In addition, if the Airy's function is independent of the angle φ, then the development of Eq. B.1 gives:

$$\frac{d^4\Phi}{dr^4} + \frac{2}{r}\frac{d^3\Phi}{dr^3} - \frac{1}{r^2}\frac{d^2\Phi}{dr^2} + \frac{1}{r^3}\frac{d\Phi}{dr} = 0 \tag{B.2}$$

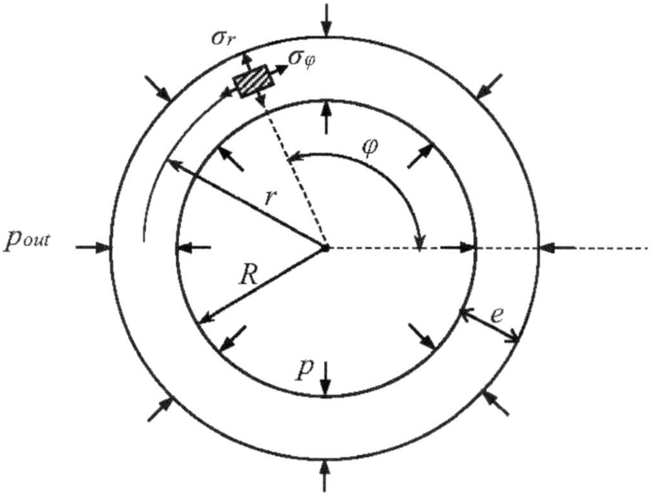

Fig. (B.1). Stress distribution in pressurized ring.

By taking $r = \exp(t)$, the general solution for the above differential equation is:

$$\phi = A \,\mathrm{Log}\, r + Br^2 \,\mathrm{Log}\, r + Cr^2 + D \tag{B.3}$$

Moreover, if the volume densities of forces are neglected, the stresses are:

$$\sigma_r = \frac{1}{r^2}\frac{\partial^2 \phi}{\partial r^2} + \frac{1}{r}\frac{\partial \phi}{\partial r} \tag{B.4}$$

$$\sigma_\varphi = \frac{\partial^2 \phi}{\partial r^2} \tag{B.5}$$

Thus, by taking account of Eq. (B.3), it follows:

$$\sigma_r = \frac{A}{r^2} + B(1 + 2\log r) + 2C \tag{B.6}$$

$$\sigma_\varphi = -\frac{A}{r^2} + B(3 + 2\log r) + 2C \tag{B.7}$$

$$\tau_{r\varphi} = 0 \tag{B.8}$$

The condition $B = 0$ resulting from the displacement condition leads to:

$$\sigma_r = \frac{A}{r^2} + 2C \tag{B.9}$$

$$\sigma_\varphi = -\frac{A}{r^2} + 2C \tag{B.10}$$

The constants of integration are obtained thanks to the boundary conditions $(\sigma_r)_{r=R} = -p$ and $(\sigma_r)_{r=R+e} = -p_{out}$, hence:

$$\frac{A}{R^2} + 2C = -p \tag{B.11}$$

$$\frac{A}{(R+e)^2}+2C=-p_{out} \tag{B.12}$$

which implies:

$$A=\frac{R^2(R+e)^2}{2\left(R+\frac{1}{2}e\right)e}(p_{out}-p) \tag{B.13}$$

$$2C=\frac{R^2p-(R+e)^2p_{out}}{2\left(R+\frac{1}{2}e\right)e}A \tag{B.14}$$

Consequently,

$$\sigma_r=\frac{1}{(2R+e)e}\left[-R^2\left(\frac{(R+e)^2}{r^2}-1\right)p+(R+e)^2\left(\frac{R^2}{r^2}-1\right)p_{out}\right] \tag{B.15}$$

$$\sigma_\varphi=\frac{1}{(2R+e)e}\left[R^2\left(\frac{(R+e)^2}{r^2}+1\right)p-(R+e)^2\left(\frac{R^2}{r^2}+1\right)p_{out}\right] \tag{B.16}$$

In case of $p_{out}=0$, Eqs. (B.15) and (B.16) become:

$$\sigma_r=\frac{R^2}{(2R+e)e}\left(1-\frac{(R+e)^2}{r^2}\right)p \tag{B.17}$$

$$\sigma_\varphi=\frac{R^2}{(2R+e)e}\left(1+\frac{(R+e)^2}{r^2}\right)p \tag{B.18}$$

Eqs. (B.17) and (B.18) show that σ_r is a compression stress and σ_φ is a strength stress that takes its maximum at the outside surface of the pipe:

$$\sigma_{\varphi.\max} = \frac{2R^2}{(2R+e)e}p \tag{B.19}$$

Appendix C: Calculation of the Classical Water Hammer Model Using the MOC

The standard reservoir-pipe-valve system is considered. Water hammer is caused by the fast valve closure. The integration of the compatibility equations (1.) leads to:

$$C_f^+ : \left.H\right|_i^j - \left.H\right|_{i-1}^{j-1} + \frac{C_f}{g}\left(\left.V\right|_i^j - \left.V\right|_{i-1}^{j-1}\right) + \frac{f\Delta z}{2gD}\left.V\right|_{i-1}^{j-1}\left|\left.V\right|_{i-1}^{j-1}\right| = 0 \tag{C.1}$$

$$C_f^- : \left.H\right|_i^j - \left.H\right|_{i+1}^{j-1} - \frac{C_f}{g}\left(\left.V\right|_i^j - \left.V\right|_{i+1}^{j-1}\right) - \frac{f\Delta z}{2gD}\left.V\right|_{i+1}^{j-1}\left|\left.V\right|_{i+1}^{j-1}\right| = 0 \tag{C.2}$$

Eq. (C.1) is valid for $dz/dt = +C_f$, whereas Eq. (C.2) is valid for $dz/dt = -C_f$. It follows:

$$\left.H\right|_i^j = \frac{1}{2}\left[\left.H\right|_{i-1}^{j-1} + \left.H\right|_{i+1}^{j-1} + \frac{C_f}{g}\left(\left.V\right|_{i-1}^{j-1} - \left.V\right|_{i+1}^{j-1}\right)\right. \\ \left. - \frac{f\Delta z}{4gD}\left(\left.V\right|_{i-1}^{j-1}\left|\left.V\right|_{i-1}^{j-1}\right| - \left.V\right|_{i+1}^{j-1}\left|\left.V\right|_{i+1}^{j-1}\right|\right)\right] \tag{C.3}$$

$$\left.V\right|_i^j = \frac{1}{2}\left[\left.V\right|_{i-1}^{j-1} + \left.V\right|_{i+1}^{j-1} + \frac{g}{C_f}\left(\left.H\right|_{i-1}^{j-1} - \left.H\right|_{i+1}^{j-1}\right)\right. \\ \left. - \frac{f\Delta t}{4D}\left(\left.V\right|_{i-1}^{j-1}\left|\left.V\right|_{i-1}^{j-1}\right| + \left.V\right|_{i+1}^{j-1}\left|\left.V\right|_{i+1}^{j-1}\right|\right)\right] \tag{C.4}$$

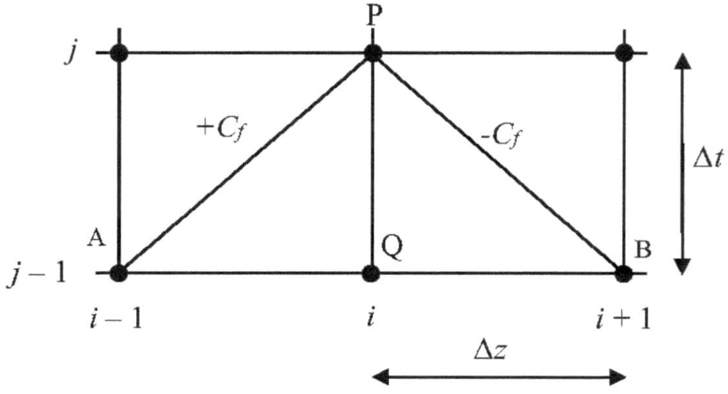

Fig. (C.1). Characteristic directions in the (z, t) plan.

Initial conditions are the steady state conditions characterized by the fluid velocity which is assumed to be uniform at each section and the pressure head given, thanks to the Darcy-Weisbach formula:

$$H_0(z) = H_{res} - \frac{fV_0^2}{2gD} z \tag{C.5}$$

with H_{res} is the constant head of the reservoir.

Boundary conditions are derived from the physical properties of the hydraulic configuration. One equation is needed at each end of the pipe. At the upstream end (constant head reservoir), one can write:

$$H\Big|_1^j = H_{res} \tag{C.6}$$

And at the downstream end (instantaneous closure valve):

$$V\Big|_{N+1}^j = 0 \tag{C.7}$$

Appendix D: Algebraic System Components for Calculation of the FSI Models

To facilitate the computer programme implementation of the FSI model (four-equation model), the numerical integration of the compatibility equations can be written in general form as follows [89]:

$$\mathbf{My} = \mathbf{k} + \mathbf{diag}(\mathbf{MY}) \tag{D.1}$$

with **M** is the coefficient matrix, **y** is the vector of unknowns, **Y** is the matrix of calculated variables and **k** is the first right-hand side vector; the second right-hand side vector is the diagonal of the matrix **MY**. Table **D.1** illustrates these mathematical items for each FSI model.

Table D.1. Algebraic system components for FSI models [89].

FSI model	Coefficient matrix			Vector y	Matrix Y	Right hand side vector		
	Inner sections	Reservoir	valve (fixed/free)			Inner sections	Reservoir	valve (fixed/free)
4EM and 4EFM	M1	R1	V1	y1	Y1	k1	r1	v1
Coupled DVCM	M2	R2	V2	y2	Y2	k2	r2	v2
Coupled DGCM	M2	R2	V2	y2	Y2	k2	r2	v2
4EVEM	M3	R3	V3	y1	Y1	k3	r3	v3
Coupled VEDVCM	M4	R4	V4	y2	Y2	k4	r4	v4

In the following expressions:

- The superscript Q denotes the one-step earlier point from the computational point P.
- For the column separation models, the piezometric head at the computational point P is $H^P = H_v - z\sin\gamma$.

It is worth noting that the expressions of matrices and vectors at boundaries are only valid for the downstream valve system; the calculation of the upstream valve system (case of cavitation in VE pipelines in this work) can be obtained similarly according to its boundary conditions.

It is also noted that some expressions used at boundary conditions such as $\hat{\gamma}_1$ and $\hat{\gamma}_2$ in the vector **r4** should be calculated separately.

Coefficient Matrices

$$\mathbf{M1} = \begin{bmatrix} 1 & K_f g & \Sigma_f & -\dfrac{\Sigma_f}{\rho_p \tilde{C}_f} \\ 1 & -K_f g & \Sigma_f & \dfrac{\Sigma_f}{\rho_p \tilde{C}_f} \\ 1 & K_p g & \Pi & -\dfrac{\tilde{C}_p}{\Sigma_p E} \\ 1 & -K_p g & \Pi & \dfrac{\tilde{C}_p}{\Sigma_p E} \end{bmatrix}; \mathbf{M2} = \begin{bmatrix} 1/A & 0 & \Sigma_f & -\dfrac{\Sigma_f}{\rho_p \tilde{C}_f} \\ 0 & 1/A & \Sigma_f & \dfrac{\Sigma_f}{\rho_p \tilde{C}_f} \\ 1/A & 0 & \Pi & -\dfrac{\tilde{C}_p}{\Sigma_p E} \\ 0 & 1/A & \Pi & \dfrac{\tilde{C}_p}{\Sigma_p E} \end{bmatrix}$$

$$\mathbf{M3} = \begin{bmatrix} 1 & K_f g - b\hat{C}_1 \dfrac{D}{2e}\rho_f g J & \Sigma_f & -\dfrac{\Sigma_f}{\rho_p \tilde{C}_f} \\ 1 & -K_f g + b\hat{C}_1 \dfrac{D}{2e}\rho_f g J & \Sigma_f & \dfrac{\Sigma_f}{\rho_p \tilde{C}_f} \\ 1 & K_p g - a\hat{C}_2 \dfrac{D}{2e}\rho_f g J & \Pi & -\dfrac{\tilde{C}_p}{\Sigma_p E} \\ 1 & -K_p g + a\hat{C}_2 \dfrac{D}{2e}\rho_f g J & \Pi & \dfrac{\tilde{C}_p}{\Sigma_p E} \end{bmatrix}; \mathbf{M4} = \begin{bmatrix} 1/A & 0 & \Sigma_f & -\dfrac{\Sigma_f}{\rho_p \tilde{C}_f} \\ 0 & 1/A & \Sigma_f & \dfrac{\Sigma_f}{\rho_p \tilde{C}_f} \\ 1/A & 0 & \Pi & -\dfrac{\tilde{C}_p}{\Sigma_p E} \\ 0 & 1/A & \Pi & \dfrac{\tilde{C}_p}{\Sigma_p E} \end{bmatrix}$$

$$\mathbf{R1} = \begin{bmatrix} 0 & 1 & 0 & 0 \\ 1 & -K_f g & \Sigma_f & \dfrac{\Sigma_f}{\rho_p \tilde{C}_f} \\ 0 & 0 & 1 & 0 \\ 1 & -K_p g & \Pi & \dfrac{\tilde{C}_p}{\Sigma_p E} \end{bmatrix} ; \ \mathbf{R2} = \begin{bmatrix} 1 & -1 & 0 & 0 \\ 0 & 1/A & \Sigma_f & \dfrac{\Sigma_f}{\rho_p \tilde{C}_f} \\ 0 & 0 & 1 & 0 \\ 0 & 1/A & \Pi & \dfrac{\tilde{C}_p}{\Sigma_p E} \end{bmatrix}$$

$$\mathbf{R3} = \begin{bmatrix} 0 & 1 & 0 & 0 \\ 1 & -K_f g + b\hat{C}_1 \dfrac{D}{2e} \rho_f g J & \Sigma_f & \dfrac{\Sigma_f}{\rho_p \tilde{C}_f} \\ 0 & 0 & 1 & 0 \\ 1 & -K_p g + a\hat{C}_2 \dfrac{D}{2e} \rho_f g J & \Pi & \dfrac{\tilde{C}_p}{\Sigma_p E} \end{bmatrix} ; \mathbf{R4} = \begin{bmatrix} 1 & -1 & 0 & 0 \\ 0 & 1/A & \Sigma_f & \dfrac{\Sigma_f}{\rho_p \tilde{C}_f} \\ 0 & 0 & 1 & 0 \\ 0 & 1/A & \Pi & \dfrac{\tilde{C}_p}{\Sigma_p E} \end{bmatrix}$$

For fixed valve:

$$\mathbf{V1} = \begin{bmatrix} 1 & K_f g & \Sigma_f & -\dfrac{\Sigma_f}{\rho_p \tilde{C}_f} \\ 1 & 0 & 0 & 0 \\ 1 & K_p g & \Pi & -\dfrac{\tilde{C}_p}{\Sigma_p E} \\ 0 & 0 & 1 & 0 \end{bmatrix} ; \ \mathbf{V2} = \begin{bmatrix} 1/A & 0 & \Sigma_f & -\dfrac{\Sigma_f}{\rho_p \tilde{C}_f} \\ 0 & 1 & 0 & 0 \\ 1/A & 0 & \Pi & -\dfrac{\tilde{C}_p}{\Sigma_p E} \\ 0 & 0 & 1 & 0 \end{bmatrix}$$

$$\mathbf{V3}=\begin{bmatrix} 1 & K_f g - b\hat{C}_1 \frac{D}{2e}\rho_f gJ & \Sigma_f & -\frac{\Sigma_f}{\rho_p \tilde{C}_f} \\ 1 & 0 & 0 & 0 \\ 1 & K_p g - a\hat{C}_2 \frac{D}{2e}\rho_f gJ & \Pi & -\frac{\tilde{C}_p}{\Sigma_p E} \\ 0 & 0 & 1 & 0 \end{bmatrix};\mathbf{V4}=\begin{bmatrix} 1/A & 0 & \Sigma_f & -\frac{\Sigma_f}{\rho_p \tilde{C}_f} \\ 0 & 1 & 0 & 0 \\ 1/A & 0 & \Pi & -\frac{\tilde{C}_p}{\Sigma_p E} \\ 0 & 0 & 1 & 0 \end{bmatrix}$$

For free valve:

$$\mathbf{V1}=\begin{bmatrix} 1 & K_f g & \Sigma_f & -\frac{\Sigma_f}{\rho_p \tilde{C}_f} \\ 1 & 0 & -1 & 0 \\ 1 & K_p g & \Pi & -\frac{\tilde{C}_p}{\Sigma_p E} \\ 0 & -A\rho_f g & \frac{m}{\Delta t}+c+\frac{k\Delta t}{2} & A_p \end{bmatrix};\mathbf{V2}=\begin{bmatrix} 1/A & 0 & \Sigma_f & -\frac{\Sigma_f}{\rho_p \tilde{C}_f} \\ 0 & 1 & -1 & 0 \\ 1/A & 0 & \Pi & -\frac{\tilde{C}_p}{\Sigma_p E} \\ 0 & 0 & \frac{m}{\Delta t}+c+\frac{k\Delta t}{2} & A_p \end{bmatrix}$$

$$\mathbf{V3}=\begin{bmatrix} 1 & K_f g - b\hat{C}_1 \frac{D}{2e}\rho_f gJ & \Sigma_f & -\frac{\Sigma_f}{\rho_p \tilde{C}_f} \\ 1 & 0 & 0 & 0 \\ 1 & K_p g - a\hat{C}_2 \frac{D}{2e}\rho_f gJ & \Pi & -\frac{\tilde{C}_p}{\Sigma_p E} \\ 0 & -A\rho_f g & \frac{m}{\Delta t}+c+\frac{k\Delta t}{2} & A_p \end{bmatrix};\mathbf{V4}=\begin{bmatrix} 1/A & 0 & \Sigma_f & -\frac{\Sigma_f}{\rho_p \tilde{C}_f} \\ 0 & 1 & -1 & 0 \\ 1/A & 0 & \Pi & -\frac{\tilde{C}_p}{\Sigma_p E} \\ 0 & 0 & \frac{m}{\Delta t}+c+\frac{k\Delta t}{2} & A_p \end{bmatrix}$$

Matrices of Calculated Variables

$$\mathbf{Y1} = \begin{bmatrix} V_z^{A_1} & V_z^{A_2} & V_z^{A_3} & V_z^{A_4} \\ H^{A_1} & H^{A_2} & H^{A_3} & H^{A_4} \\ \dot{u}_z^{A_1} & \dot{u}_z^{A_2} & \dot{u}_z^{A_3} & \dot{u}_z^{A_4} \\ \sigma_z^{A_1} & \sigma_z^{A_2} & \sigma_z^{A_3} & \sigma_z^{A_4} \end{bmatrix} ; \ \mathbf{Y2} = \begin{bmatrix} Q_u^{A_1} & Q_u^{A_2} & Q_u^{A_3} & Q_u^{A_4} \\ Q_d^{A_1} & Q_d^{A_2} & Q_d^{A_3} & Q_d^{A_4} \\ \dot{u}_z^{A_1} & \dot{u}_z^{A_2} & \dot{u}_z^{A_3} & \dot{u}_z^{A_4} \\ \sigma_z^{A_1} & \sigma_z^{A_2} & \sigma_z^{A_3} & \sigma_z^{A_4} \end{bmatrix}$$

Vectors of Unknowns

$$\mathbf{y1} = \begin{bmatrix} V_z \\ H \\ \dot{u}_z \\ \sigma_z \end{bmatrix} ; \ \mathbf{y2} = \begin{bmatrix} Q_u \\ Q_d \\ \dot{u}_z \\ \sigma_z \end{bmatrix}$$

First Right-hand Side Vectors

$$\mathbf{k1} = \begin{bmatrix} b\Delta tg\left[\left(\Gamma\Sigma_f - 1\right)h_f^{A_1} + \Sigma_f \sin\gamma\right] \\ b\Delta tg\left[\left(\Gamma\Sigma_f - 1\right)h_f^{A_2} + \Sigma_f \sin\gamma\right] \\ a\Delta tg\left[\left(\Gamma\Pi - 1\right)h_f^{A_3} + \Pi \sin\gamma\right] \\ a\Delta tg\left[\left(\Gamma\Pi - 1\right)h_f^{A_4} + \Pi \sin\gamma\right] \end{bmatrix} ; \ \mathbf{k2} = \mathbf{k1} + \begin{bmatrix} -K_f g\left(H^P - H^{A_1}\right) \\ K_f g\left(H^P - H^{A_2}\right) \\ -K_p g\left(H^P - H^{A_3}\right) \\ K_p g\left(H^P - H^{A_4}\right) \end{bmatrix} ;$$

$$\mathbf{k3} = \mathbf{k1} + \begin{bmatrix} b\Delta t\hat{\gamma}_1 \\ -b\Delta t\hat{\gamma}_1 \\ a\Delta t\hat{\gamma}_2 \\ -a\Delta t\hat{\gamma}_2 \end{bmatrix}; \ \mathbf{k4} = \mathbf{k1} + \begin{bmatrix} b\Delta t\hat{\gamma}_1 - K_f g\left(H^P - H^{A_1}\right) + b\Delta t\hat{C}_1 \dfrac{\partial\left(\varepsilon_{\varphi r}\right)^P}{\partial t} \\ -b\Delta t\hat{\gamma}_1 + K_f g\left(H^P - H^{A_2}\right) - b\Delta t\hat{C}_1 \dfrac{\partial\left(\varepsilon_{\varphi r}\right)^P}{\partial t} \\ a\Delta t\hat{\gamma}_2 - K_p g\left(H^P - H^{A_3}\right) + a\Delta t\hat{C}_2 \dfrac{\partial\left(\varepsilon_{\varphi r}\right)^P}{\partial t} \\ -a\Delta t\hat{\gamma}_2 + K_p g\left(H^P - H^{A_4}\right) - a\Delta t\hat{C}_2 \dfrac{\partial\left(\varepsilon_{\varphi r}\right)^P}{\partial t} \end{bmatrix}$$

$$\mathbf{r1} = \begin{bmatrix} H_{res} \\ b\Delta tg\left[\left(\Gamma\Sigma_f - 1\right)h_f^{A_2} + \Sigma_f \sin\gamma\right] \\ 0 \\ a\Delta tg\left[\left(\Gamma\Pi - 1\right)h_f^{A_4} + \Pi\sin\gamma\right] \end{bmatrix}; \ \mathbf{r2} = \mathbf{r1} + \begin{bmatrix} 0 \\ K_f g\left(H^P - H^{A_2}\right) \\ 0 \\ K_p g\left(H^P - H^{A_4}\right) \end{bmatrix}$$

$$\mathbf{r3} = \mathbf{r1} + \begin{bmatrix} 0 \\ -b\Delta t\hat{\gamma}_1 \\ 0 \\ -a\Delta t\hat{\gamma}_2 \end{bmatrix}; \ \mathbf{r4} = \mathbf{r1} + \begin{bmatrix} 0 \\ -b\Delta t\hat{\gamma}_1 + K_f g\left(H^P - H^{A_2}\right) - b\Delta t\hat{C}_1 \dfrac{\partial\left(\varepsilon_{\varphi r}\right)^P}{\partial t} \\ 0 \\ -a\Delta t\hat{\gamma}_2 + K_p g\left(H^P - H^{A_4}\right) - a\Delta t\hat{C}_2 \dfrac{\partial\left(\varepsilon_{\varphi r}\right)^P}{\partial t} \end{bmatrix}$$

For fixed valve:

$$\mathbf{v1}=\begin{bmatrix} b\Delta tg\left[\left(\Gamma\Sigma_f-1\right)h_f^{A_1}+\Sigma_f\sin\gamma\right] \\ 0 \\ a\Delta tg\left[\left(\Gamma\Pi-1\right)h_f^{A_3}+\Pi\sin\gamma\right] \\ 0 \end{bmatrix};\mathbf{v2}=\mathbf{v1}+\begin{bmatrix} -K_f g\left(H^P-H^{A_1}\right) \\ 0 \\ -K_p g\left(H^P-H^{A_3}\right) \\ 0 \end{bmatrix}$$

$$\mathbf{v3}=\mathbf{v1}+\begin{bmatrix} b\Delta t\hat{\gamma}_1 \\ 0 \\ a\Delta t\hat{\gamma}_2 \\ 0 \end{bmatrix};\quad \mathbf{v4}=\mathbf{v1}+\begin{bmatrix} b\Delta t\hat{\gamma}_1-K_f g\left(H^P-H^{A_1}\right)+b\Delta t\hat{C}_1\dfrac{\partial\left(\varepsilon_{\varphi r}\right)^P}{\partial t} \\ 0 \\ a\Delta t\hat{\gamma}_2-K_p g\left(H^P-H^{A_3}\right)+a\Delta t\hat{C}_2\dfrac{\partial\left(\varepsilon_{\varphi r}\right)^P}{\partial t} \\ 0 \end{bmatrix}$$

For free valve:

$$\mathbf{v1}=\begin{bmatrix} b\Delta tg\left[\left(\Gamma\Sigma_f-1\right)h_f^{A_1}+\Sigma_f\sin\gamma\right] \\ 0 \\ a\Delta tg\left[\left(\Gamma\Pi-1\right)h_f^{A_3}+\Pi\sin\gamma\right] \\ \left(\dfrac{m}{\Delta t}-\dfrac{k\Delta t}{2}\right)\dot{u}_z^Q-ku_z^Q+A\rho_f gL\sin\gamma \end{bmatrix};\mathbf{v2}=\mathbf{v1}+\begin{bmatrix} -K_f g\left(H^P-H^{A_1}\right) \\ 0 \\ -K_p g\left(H^P-H^{A_3}\right) \\ 0 \end{bmatrix}$$

$$\mathbf{v3} = \mathbf{v1} + \begin{bmatrix} b\Delta t\hat{\gamma}_1 \\ 0 \\ a\Delta t\hat{\gamma}_2 \\ 0 \end{bmatrix};$$

Appendix E: Frequency Dependent Friction

The unsteady friction approach consists in incorporating an additional term into the momentum equation in order to further consideration of the frictional and the inertial effects of the non-uniform velocity profile. The slope of the energy line h_f is decomposed respectively into two terms: steady-state term and unsteady term.

$$h_f = h_{f.s} + h_{f.u} \quad \textbf{(E.1)}$$

Unsteady friction models derive from the extra losses caused by the two-dimensional nature of the unsteady velocity profile [90]. Several types of unsteady friction models exist in the literature and are described in [91]. By assuming the velocity V is uniform on each section A, Zielke defined the head loss h_f as a sum of a steady-state part defined by the Darcy-Weisbach formulae and an unsteady part [90].

$$h_f = \frac{f}{2gD} V|V| + \frac{16v'}{gD^2}\left(\frac{\partial V}{\partial t} * W\right)(t) \quad \textbf{(E.2)}$$

where W is the weighting function, in time, v' is the kinematic viscosity and (*) denotes convolution. The unsteady term follows from the convolution of the weighting function W with past temporal velocity variations.

$$\left(\frac{\partial V}{\partial t} * W\right)(t) = \int_0^t \frac{\partial V}{\partial t}(u) \cdot W(t-u)\,du \quad \textbf{(E.3)}$$

The convolution-based unsteady frictional head loss term in a staggered grid (SG), called *full convolution scheme* is [92]:

$$h_f(z,t) = \frac{fV(z,t)|V(z,t)|}{2gD} +$$

$$\frac{16\nu'}{gD^2} \sum_{j=1,3,5,\ldots}^{M} \left[V(z,t-j\Delta t+\Delta t) - V(z,t-j\Delta t-\Delta t)\right] W(j\Delta t) \tag{E.4}$$

With $M = t/\Delta t - 1$. The weighting functions are defined in terms of the dimensionless time $\tau = 4\nu' t/D^2$ [92]. For laminar flow, Zielke has developed the weighting functions by assuming a constant kinematic viscosity [53, 59, 89,90-92]. For turbulent flow, the Vardy-Brown's model is obtained by deriving weighting functions for smooth pipes and rough pipes by using a linearly varied frozen turbulent viscosity in the shear-layer and infinite viscosity in the core [92].

$$W(\tau) = \frac{A^* \exp(-B^* \tau)}{\sqrt{\tau}} \tag{E.5}$$

with A^* and B^* are fitted coefficients to a more complex theoretical weighting function. For smooth pipes, the coefficients are:

$$A^* = \frac{1}{2}\sqrt{\frac{\nu'_w}{\pi \nu'_{lam}}}, \quad B^* = \frac{\mathrm{Re}^{\kappa}}{12.86} \quad \text{and} \quad \kappa = \log\left(15.29\,\mathrm{Re}^{-0.0567}\right) \tag{E.6}$$

where ν'_{lam} is the laminar kinematic viscosity and ν'_w is the kinematic viscosity at the pipe wall. For fully rough-pipe turbulent flow, the Vardy-Brown's coefficients become:

$$A^* = 0.0103\sqrt{\mathrm{Re}}\left(\frac{\varepsilon}{D}\right)^{0.39} \quad \text{and} \quad B^* = 0.0352\,\mathrm{Re}\left(\frac{\varepsilon}{D}\right)^{0.41} \tag{E.7}$$

in which, the relative roughness ε/D is in the range from 10^{-6} to 10^{-2}.

BIBLIOGRAPHY

[1] D.I.C. Covas, "Inverse transient analysis for leak detection and calibration of water pipe systems modelling special dynamic effects", In: PhD Thesis, Department of Civil and Environmental Engineering, Imperial College of Science, Technology and Medecine. London, UK, 2003.
[2] A.S. Tijsseling, "Fluid structure interaction in case of waterhammer with cavitation", In: Ph.D. Thesis, Delft University of Technology, Faculty of Civil Engineering, Communication on Hydraulic and Geotechnical Engineering. Delft, Netherlands, 1993, pp. 93-96.
[3] L-F. Ménabréa, "Note on the effect of water impact on pipes", Compte Rendus Hebdomadaires des Sciences de l'Académie des Sciences, Paris, France, vol. I, pp. 221-224, 1958.
[4] L-F. Ménabréa, "Note on the effect of water impact on pipes", Annals of Civil Engineering, Paris, France, vol. I, pp. 269-275, 1962.
[5] N. Joukowsky, "Hydraulic water hammer in water filled pipelines", In: Memory of the Scientific Imperial Academy of St-Petersbourg., vol. 9. Germany, 1900.
[6] E.B. Wylie, and V.L. Streeter, Fluid transients. McGraw-Hill: New York, USA, 1993.
[7] W. Weber, "Wave propagation theory in water or other incompressible liquid in elastic pipes", Mathematical-Physical Section, Leipzig, Germany, vol. 18, pp. 353-357, 1866.
[8] A.I. Moens, The Pulsation. E.J. Brill. Leydn, Hollande, 1878.
[9] D.J. Korteweg, "Sound speed in elastic pipes", Annals of Physics and Chemestry, vol. 5, no. 12, pp. 542-525, 1878.
[10] J. Parmakian, Waterhammer analysis. Prentice-Hall: New York, US, 1955.
[11] V.L. Streeter, and E.B. Wylie, Hydraulic transients. McGraw-Hill: New York, USA, 1967.
[12] I.S. Gromeka, "On the velocity of propagation of wave-like motion of fluids in elastic tubes". Physical-Mathematical Section of the Scientific Society of the Imperial University of Kazan: Kazan, Russia, 1883, pp. 1-19. (in Russian)
[13] H. Lamb, "On the velocity of sound in a tube, as affected by the elasticity of the walls", Memoirs of the Manchester Literary and Philosophical Society, vol. 42, Manchester, UK, pp. 1-16, 1898.
[14] R. Skalak, "An extended of the theory of waterhammer", Trans of the ASME, vol. 78, no. 1, pp. 105-116, 1956.
[15] A.R.D. Thorley, "Pressure transients in hydraulic pipelines", *J. Basic Eng.*, vol. 91, no. 3, pp. 453-460, 1969.
[http://dx.doi.org/10.1115/1.3571152]
[16] A.S. Tijsseling, A.E. Vardy, and D. Fan, "Fluid-structure interaction and cavitation in a single-elbow pipe system", *J. Fluids Struct.*, vol. 10, no. 4, pp. 395-420, 1996.
http://dx.doi.org/10.1006/jfls.1996.0025
[17] A.G.T.J Heinsbroek, "Fluid-structure interaction in non-rigid pipeline systems, Nuclear Engineering and Design", Delft Hydraulics, no. 172, pp. 123-135, 1997.
[18] G.B. Wallis, "One-dimensional two-phase flow". McGraw-Hill: New York, US, 1969.
[19] F.R. Young, Cavitation. McGraw-Hill: London, UK, 1989.
[20] E. Haj Taïeb, "Transient flow in deformable pipes with vapour cavitation and dissolved air release", In: Thesis report, Faculty of Sciences of Tunis. Tunisia, 1999.

Abdelaziz Ghodhbani, Ezzeddine Haj Taïeb, Mohsen Akrout & Sami Elaoud

[21] F. Caupin, and E. Herbert, "Cavitation in water: A review", laboratory of statistic physics of the higher normal school. French: Paris, pp. 1000-1017, 2006.
[22] A.Bergant, A. Tijsseling, J. Vitkovsky, and A. Simpson, "Discrete vapour cavity model with improved timing of opening and collapse of cavities", Proc. 2nd IAHR International Meeting of the Workgroup on Cavitation and Dynamic Problems in Hydraulic Machinery and Systems, Timişoara, Romania pp. 117-128, 2007.
[23] A.R.D. Thorley, "Fluid transients in pipeline systems", In: Thermo-Fluids Engineering Research Center, 1st City University: London EC1V 0HB, UK, 1991.
[24] A.R. Simpson, "Large water hammer pressures due to column separation in sloping pipes", In: Thermo-Fluids Engineering Research Center, 1st City University: London EC1V 0HB, UK, 1986.
[25] A. Bergant, J. Vitkovsky, A. Simpson, M. Lambert, and A. Tijsseling, "Discrete vapour cavity model with efficient and accurate unsteady friction term", Department of mathematics and computer science, Yokohama, Japan, 2006.
[26] T.H. Hogg, and J.J. Trail, "Discussion of Speed changes of hydraulic turbine for sudden changes of load by E.B. Strowger & S.L. Kerr"., Trans. Of the ASME, vol. 48, pp. 252-257, 1926.
[27] A. Langevin, "Bulletin of the technical building union", *France*, 1928.
[28] J.N. LeConte, "Experiments and calculations on the resurge phase of water hammer", *Trans. Am. Soc. Mech. Eng*., vol. 59, no. 8, pp. 691-694, 1937. [http://dx.doi.org/10.1115/1.4020577]
[29] H.R. Lupton, "Graphical analysis of pressure surges in piping systems", Trans of the ASME, vol. 59, pp. 691-694, 1953.
[30] I.C. O'Neill, "Water-hammer in simple pipe systems", In: MSc Thesis, University of Melbourne. Melbourne, Australia, 1959.
[31] B.B. Sharp, "Cavity formation in simple pipes due to rupture of water column", Nature, vol. 185, no. 4709, pp. 302-303, 1960. [http://dx.doi.org/10.1038/185302b0]
[32] B.B. Sharp, "The growth and collapse of cavities produced by a rarefaction wave with particular reference to rupture of water column", In: PhD. Thesis, The university of Melbourne. Melbourne, Australia, 1965.
[33] B.B. Sharp, "Rupture of water column", Proceeding of the 2nd Australasian Conf. on Hydraulics and Fluid Mechanics, Auckland, New Zealand pp. A169-A176, 1965.
[34] V. Jordan, "The Influence of check valves on water hammer at pump failure", Strojniski Vestnik, vol. 7, no. 4, 5, pp. 19-21, 1961.
[35] A.R. Simpson, and E.B. Wylie, "Towards an improved understanding of waterhammer column separation in pipelines,Civil Engineering Transactions 1989", Institution of Engineers, Australia, vol. CE31, no. 3, pp. 113-120, 1989.
[36] J.J. Shu, "Modelling vaporous cavitation on fluid transients", international journal of pressure vessels and piping, school of mechanical and production engineering, nanyang technological university, 50 nanyang avenue., no. 80, pp. 187-195, 2003.
[37] F. Knapp, "Discussion of experiments and calculations on the resurge phase of water hammer", Transactions of the ASME 61, pp. 440-441, 1939.
[38] V. Jordan, "Prediction of water hammer at pump failure without surge protection under water column separation conditions", In: PhD Thesis, University of Belgrade. Belgrade, Yugoslavia, 1965.
[39] G. Pezzinga, and D. Cannizzaro, "Analysis of transient vaporous cavitation in pipes by a distributed 2D model", J. Hydraul. Eng., vol. 140, no. 6, p. 04014019, 2014.

[http://dx.doi.org/10.1061/(ASCE)HY.1943-7900.0000840]

[40] W. Zielke, and H.D. Perko, "Low-pressure phenomena and water hammer analysis", 3R international, vol. 24, pp. 348-355, 1985.

[41] S.W. Kieffer, "Sound speeds in liquid-gas mixtures: Water-air and water-steam", JGR, vol. 182, no. 20, pp. 2895-2904, 1977. [http://dx.doi.org/10.1029/JB082i020p02895]

[42] R.W. Angus, "Water hammer in pipes, including those supplied by centrifugal pumps: Graphical treatment", *Proc.- Inst. Mech. Eng.*, vol. 136, no. 1, pp. 245-331, 1937. [http://dx.doi.org/10.1243/PIME_PROC_1937_136_021_02]

[43] A.Bergant, Transient cavitating flow in pipelines. PhD Thesis, University of Ljubljana: Ljubljana, Slovenia, 1992.

[44] A.E. Vardy, and D. Fan, "Flexural waves in a closed tube", Proc. of the 5th Int. Conf. on Pressure Surges, pp. 43-47, 1989.

[45] D.H. Wilkinson, and E.M. Curtis, "Waterhammer in a thin-walled pipe", Proc. of the 3rd Int. Conf. on Pressure Surges, pp. 221-2240, 1980.

[46] A.S. Tijsseling, "Skalak's extended theory of water hammer", In: Journal of Sound and Vibration, Department of Mathematics and Computer Science. University of Technology of Eindhoven: Eindhoven, The Netherlands, no. 310, pp. 718-728, 2007.

[47] L. Hadj-Taïeb, "Vapour cavitation in transient flow with fluid-structure interaction", In: Thesis report. National Engineering School of Sfax: Sfax, Tunisia, 2008.

[48] B.B. Sharp, "A simple model for water column rupture", In: Proceeding of the 17th IAHR Congress, Baden-Baden, Germany no. 5, pp. 155-161, 1977.

[49] A.S. Tijsseling, and C.S.W. Lavooij, "Waterhammer with fluid-structure interaction", *Appl. Sci. Res.*, vol. 47, no. 3, pp. 273-285, 1990. [http://dx.doi.org/10.1007/BF00418055]

[50] D. Fan, and A. Tijsseling, "Fluid-structure interaction with cavitation in transient pipe flows", *Fluids Eng.*, vol. 114, no. 2, pp. 268-274, 1992. [http://dx.doi.org/10.1115/1.2910026]

[51] A.S. Tijsseling, "Fluid structure interaction in liquid-filled pipe systems: A review", *J. Fluids Struct.*, vol. 10, no. 2, pp. 109-146, 1996. [http://dx.doi.org/10.1006/jfls.1996.0009]

[52] A.Ghodhbani, M. Akrout, and E. Haj Taïeb, "Coupled approach and calculation of the discrete vapour cavity model", In: Journal of Fluids and Structures. Department of Civil Engineering, University of Dundee: Dundee, UK, no. 10, pp. 109-146, 2019. [http://dx.doi.org/10.1016/j.jfluidstructs.2019.102691]

[53] A.Bergant, and A.R. Simpson, "Estimating unsteady friction in transient cavitating pipe flow", 2nd International Conference on Water Pipeline Systems, Edinburg, Scotland pp. 3-15, 1994.

[54] V.O.P. Ostrada, "Investigation on the effects of entrained air in pipelines". Dissertation, University of Stuttgart, Heft 158: Stuttgart, Germany, 2007.

[55] A.H. De Vries, Cavitation due to waterhammer in horizontal pipelines with several high points. Delft, The Netherlands, 1973.Delft Hydraulics Laboratory, Report M 1152. Delft, The Netherlands, 1973.

[56] G.A. Provoost, "Investigation into cavitation in a prototype pipeline caused by water hammer", Proc. of the Second International Conference on Pressure Surges, pp. 13-29, 1976.

[57] G.A. Provoost, and E.B. Wylie, "Discrete gas model to represent distributed free gas in liquids", In: Proceedings of the 5th International Symposium on Water Column Separation, IAHR, Obernach, Germany. Also: Delft Hydraulics Laboratory, Publication No. 263, 1982.

[58] E.B. Wylie, "Simulation of vaporous and gaseous cavitation", *J. Fluids Eng.*, vol. 106, no. 3, pp. 307-311, 1984. [http://dx.doi.org/10.1115/1.3243120]

[59] A.Ghodhbani, and E. Haj Taïeb, "A four-equation friction model for water hammer calculation in quasi-rigid pipelines", In: Int. Journal of Pressure Vessels and Piping, Department of Mechanical Engineering, National Engineering School of Sfax: Sfax, Tunisia, pp. 54-62, 2017.

[60] J.P.T. Kalkwijk, and C. Kranenburg, "Cavitation in horizontal pipelines due to water hammer", *J. Hydraul. Div.*, vol. 97, no. 10, pp. 1585-1605, 1971. [http://dx.doi.org/10.1061/JYCEAJ.0003106]

[61] J.P.T. Kalkwijk, and C. Kranenburg, "Closure to cavitation in horizontal pipelines due to water hammer", *J. Hydraul. Div.*, vol. 99, no. 3, pp. 529-530, 1973. [http://dx.doi.org/10.1061/JYCEAJ.0003599]

[62] C. Kranenburg, "Gas release during transient cavitation in pipes", *J. Hydra. Div.*, vol. 100, no. 10, 1974. [http://dx.doi.org/10.1061/JYCEAJ.0004077]

[63] C. Kranenburg, "Transient cavitation in pipelines", PhD Thesis, Delft University of Technology, Dept. of Civil Engineering, Laboratory of Fluid Mechanics, Delft, The Netherlands. Also: Communications on Hydraulics Delft University of Technology, Dept. of Civil Engineering, Report No. 73-2, 1973, 1974.

[64] C. Kranenburg, "The effect of free gas on cavitation in pipelines induced by water hammer", In: Proceedings of the First International Conference on Pressure Surges, pp. 41-52, 1972.

[65] L. Haj-Taïeb, and E. Hadj-Taïeb, "Effect of pipe-wall viscoelasticity on vapour cavitation in transient flow", In: J. Theoretical and Applied mechanics. National Engineering School of Sfax: Sfax, Tunisia, 2007.

[66] A.Bergant, and A.R. Simpson, "Water hammer and column separation measurements in an experimental apparatus". Dept. of Civil and Environmental Engineering: University of Adelaide, Adelaide, Australia, 1995.

[67] D.I.C Covas, "The dynamic effect of pipe wall viscoelasticity in hydraulic transients. Part I-experimental analysis and creep characterization", In: Department of Civil and Environmental Engineering, Imperial College of Science, Technology and Medecine. London, UK, 2004.

[68] A.K. Soares, D.I.C. Covas, H.M. Ramos, and L.F.R. Reis, "Unsteady flow with cavitation in viscoelastic pipes", *Int. J. Fluid Mach. Syst.*, vol. 2, no. 4, pp. 269-277, 2009. [http://dx.doi.org/10.5293/IJFMS.2009.2.4.269]

[69] C.S.W. Lavooij, and A.S. Tusseling, "Fluid-structure interaction in liquid-filled piping systems", *J. Fluids Struct.*, vol. 5, no. 5, pp. 573-595, 1991. [http://dx.doi.org/10.1016/S0889-9746(05)80006-4]

[70] J. Coirier, Continuum mechanics., 2nd Ed. Higher National School of Mechanics and Aerautechnics (ENSMA) of Poitier: Paris, French, 2001.

[71] A.Keramat, A. Tijsseling, Q. Hou, and A. Ahmadi, "Fluid-structure interaction with pipe-line viscoelasticity during water hammer", In: Journal of Fluids and Structures. Department of Civil Engineering, University of Technology of Shahrood: Shahrood, Iran, 2012, no. 28, pp. 434-455.

[72] R. Zanganeh, A. Ahmadi, and A. Keramat, "Fluid-structure interaction with viscoelastic supports during waterhammer in pipelines", Journal of Fluids and Structures, vol. 54, Department of Civil Engineering, University of Technology of Shahrood: Shahrood, Iran, pp. 215-311, 2014.

[73] D.I.C. Covas, I. Soianov, J.F. Mano, H. Ramos, and N. Graham, The dynamic effect of pipe wall viscoelasticity in hydraulic transients. Part I-model development, calibration and

verification. Department of Civil and Environmental Engineering, Imperial College of Science, Technology and Medecine: London, UK, 2005.
[http://dx.doi.org/10.1080/00221680509500111]

[74] S. Henclik, Analytical solution and numerical study on water hammer in a pipeline closed with an elastically attached valve. Journal of sound and vibration, Department of Hydropower, Institute of Fluid-flow Machinery, Polish Academy of Sciences: Gdansk, Poland, 417 (2018), pp. 245-259, 2018.
[http://dx.doi.org/10.1016/j.jsv.2017.12.011]

[75] A.Ghodhbani, E. Haj Taïeb, and M. Akrout, "Effect of anchor conditions on structural responses during fluid transients in pipelines", Proceedings of the Seventh International Conference on Advances in Mechanical Engineering and Mechanics (ICAMEM), Hammamet, Tunisia 2019.

[76] D.D. Budney, D.C. Wiggert, and F.J. Hatfield, "The influence of structural damping on internal pressure during a transient pipe flow", Transactions of the ASME, *J. Fluids Eng.*, vol 113, pp. 424-429, 1991.
[http://dx.doi.org/10.1115/1.2909513]

[77] A.Bergant, and A.S. Tijsseling, Parameters affecting water hammer wave attenuation, shape and timing. Litostroj E.I. d.o.o, 1000: Ljubljana, Slovenia, 2002.

[78] Ghodhbani, and E. Haj Taïeb, "Numerical coupled modelling of water hammer in quasi-rigid thin pipes", Proceedings of the Fifth International Conference Design and Modeling of Mechanical Systems (CMSM), Tunisia pp. 253-264, 2013.
[http://dx.doi.org/10.1007/978-3-642-37143-1_31]

[79] A.R. Simpson, and A. Bergant, "Numerical comparison of pipe-column-separation models", *J. Hydraul. Eng.*, vol. 120, no. 3, pp. 361-377, 1994.
[http://dx.doi.org/10.1061/(ASCE)0733-9429(1994)120:3(361)]

[80] A.S. Tijsseling, "Exact solution of linear hyperbolic four-equation system in axial liquid-pipe vibration", Journal of Fluids and Structures, vol. 18, no. 2, pp. 179-196, 2003.
[http://dx.doi.org/10.1016/j.jfluidstructs.2003.07.001]

[81] A.S. Tijsseling, Poisson-coupling beat in extended waterhammer theory.ASME Flow-induced vibration and noise., Department of Civil Engineering, University of Dundee: Dundee, United Kingdom, Vol. 53-2, pp. 529-532, 1997.

[82] A.Bergant, and A.R. Simpson, "Pipeline column separation flow regimes", *J. Hydraul. Eng.*, vol. 125, no. 8, pp. 835-848, 1999.
[http://dx.doi.org/10.1061/(ASCE)0733-9429(1999)125:8(835)]

[83] W. Wagner, and H.J. Kretzschmar, International steam tables., 2nd ed. Springer: Germany, 2008.
http://dx.doi.org/10.1007/978-3-540-74234-0

[84] D.C. Wiggert, F.J. Hatfield, and S. Stuckenbruck, "Analysis of liquid and structural transients by the method of characteristics", *J. Fluids Eng.*, vol. 109, no. 2, pp. 161-165, 1987.
[http://dx.doi.org/10.1115/1.3242638]

[85] A.G.T.J. Heinsbroak, and A.S. Tijsseling, "The influence of support rigidity on water hammer pressures and pipe stresses", In: Proc. 2nd Int. Confer. Water Pipeline Systems: Edinburgh, UK, 1994, pp. 17-30, 1994.

[86] S. Kaneko, Flow-induced vibrations classifications and lessons from practical experiences., 2nd ed Elsevier, 2014.

[87] A.E. Vardy, and JM.B. Brown, "Transient turbulent friction in smooth pipe flows", In: Journal of Sound and Vibration, Civil Engineering Department, University of Dundee, Dundee DD1 4HN, Elsvier: Scotland, UK, 259(5), pp. 1011-1036, 2003.

http://dx.doi.org/10.1006/jsvi.2002.5160

[88] A.E. Vardy, and JM.B. Brown, "Transient turbulent friction in fully-rough pipe flows", In: Journal of Sound and Vibration Civil Engineering Department, School of Engineering, University of Dundee, Elsvier: Dundee DD1 4HN, UK, 270, pp. 233-257, 2004. [http://dx.doi.org/10.1016/S0022-460X(03)00492-9]

[89] A.Ghodhbani, "Mcdelling and Simulation of transient cavitation in pipelines", In: Thesis report, National Engineering School of Sfax, Tunisia, 2021.

[90] W. Zielke, "Frequency-dependent friction in transient pipe flow", ASME J. Basic Eng, series D, vol. 91, no. 4, 1969.

[91] A.Adamkowski, and M. Lewandowski, "Experimental examination of unsteady friction models for transient pipe flow simulation", *J. Fluids Eng.*, vol. 128, no. 6, pp. 1351-1363, 2006. [http://dx.doi.org/10.1115/1.2354521]

[92] J. Vítkovský, "Efficient and accurate calculation of zielke and vardy-brown unsteady friction in pipe transients", S.J. Murray, Ed., Proceedings of the 9th International Conference on Pressure Surges, pp. 405-419, 2004.

SUBJECT INDEX

A

B

C

D

Abdelaziz Ghodhbani, Ezzeddine Haj Taïeb, Mohsen Akrout & Sami Elaoud

I

J

K

L

M

N

O

P

R

S

T

V

W

www.ingramcontent.com/pod-product-compliance
Lightning Source LLC
Chambersburg PA
CBHW050835220326
41598CB00006B/367
* 9 7 8 9 8 1 5 1 2 3 7 8 4 *